ECONOMICS FOR FISHERIES MANAGEMENT

Our book is dedicated to the special people who share our lives: Ariana, Brecon and Carol-Anne; Haleh, Phillip, and Shirin.

Economics for Fisheries Management

R. QUENTIN GRAFTON
The Australian National University, Australia

JAMES KIRKLEY
Virginia Institute of Marine Sciences, USA

TOM KOMPAS
The Australian National University, Australia

DALE SQUIRES
National Marine Fisheries Service, and
University of California, San Diego, USA

Routledge
Taylor & Francis Group

LONDON AND NEW YORK

First published 2006 by Ashgate Publishing

2 Park Square, Milton Park, Abingdon, Oxon OX14 4RN
711 Third Avenue, New York, NY 10017, USA

Routledge is an imprint of the Taylor & Francis Group, an informa business

First issued in paperback 2016

British Library Cataloguing in Publication Data
Economics for fisheries management. - (Ashgate studies in
   environmental and natural resource economics)
  1. Fisheries - Economics
  I. Grafton, R. Quentin, 1962-
  338.3'727

Library of Congress Cataloging-in-Publication Data
Economics for fisheries management / by R. Quentin Grafton ... [et al.].
     p. cm. -- (Ashgate studies in environmental and natural resource
economics)
  Includes bibliographical references and index.
  ISBN 0-7546-3249-0
  1. Fisheries--Economic aspects. 2. Fishery management. I. Grafton, R. Quentin,
1962- II. Series.

  SH334.E24 2006
  338.3'727--dc22

2005032389

ISBN 978-0-7546-3249-8 (hbk)
ISBN 978-1-138-25209-7 (pbk)

Contents

List of Figures

List of Tables

Foreword

It is well known that on a global scale marine fisheries are in a state of crisis. Total world catches are now, for the first time, in decline, as more and more stocks become over-fished. This may be understandable for fish populations on the high seas, for which the control of exploitation seems almost impossible. But in fact the main losses have occurred within the 200-mile fishing zones that were established around 25 years ago. To put it bluntly, coastal fish stocks have been hammered, sometimes obliterated, *in spite of intensive management attempts.* The famous cod stocks of the Northwest Atlantic, which supplied much of Europe and the New World with a major source of protein for many centuries, are only the most visible example.

What on earth went wrong? Did these populations just disappear from natural causes, as some would have us believe? Too many seals, as one Newfoundland fishing company owner told me? No, the evidence is overwhelming: the stocks were over-fished, period. Also, in some cases, sea-floor habitat has been destroyed by trawling and similar operations. But shouldn't the managers have foreseen these events? In fact, it's not easy to figure out, from the data, what's happening to fish stocks at a given time. Models do make predictions, but these have often proved to be incorrect. Over-confidence in models of population dynamics has certainly been one aspect of the fisheries crisis.

This book, however, identifies another major driving force, namely *the failure of fishery managers to understand the economically motivated behavior of fishers.* To quote from Chapter 1, 'We contend that the biggest barrier to sustaining fisheries is not our ignorance of marine ecosystems, nor is it a lack of goodwill from those involved in fishing. Rather, the stumbling block to better fisheries management is a profound lack of understanding of the behavior of the oceans' greatest predator – the commercial fisher'. For example, when faced with a TAC (Total Allowable Catch quota), fishing fleets have invariably expanded as a result of their owners' desire to catch fish as quickly as possible. This has forced the managers to shorten the fishing season, but this only incites further expansion of fishing power. Over-capacity of fishing fleets worldwide is now considered itself to have reached crisis proportions. Fishery openings of a few weeks, or even a few days, per year have become commonplace. The description 'too many boats chasing too few fish' surely fits most marine fisheries today.

What could be done to improve on this 'tragedy of the commons' in marine fisheries? That is the theme of this book. Without giving away secrets, I can say that a change in property rights, for example by introducing Individual Fishing Quotas (IFQs), has the potential for reversing the situation, if the system of IFQs is properly set up and controlled. The book goes into this and other vital questions, using the standard tools of economic analysis. The authors also explain in detail how to determine the economic success (or failure) of a given management system from fishery economic data. They describe as examples eight fisheries from different parts of the world.

Economics for Fisheries Management is therefore an important and timely contribution to the fisheries literature. We are currently in a phase of rapid revision of traditional management systems; unless the messages of these four eminent Economists are taken into account, it seems unlikely that the oceans will much longer supply the world with an abundant source of tasty and nutritious food.

Colin W. Clark
Emeritus Professor
Department of Mathematics and Institute of Applied Mathematics
University of British Columbia

Preface

Economics for Fisheries Management provides you with the basic theory, and especially the tools, to understand, analyze and improve the management of marine fisheries. The book assumes no prior knowledge of economics and is ideal for researchers and students in a range of disciplines who are interested in fisheries and fisheries management. It will also be of considerable value to fisheries scientists, managers and policy makers who desire to learn more about the economics of fishing and wish to improve the state of their fisheries.

We focus on the methods and policy insights rather than the underlying theory so that the book can be of the greatest possible practical use. Thus our book is *not* about economic theory, dynamics and optimal control, subjects that are examined well in other books, but rather it explains the methods of analysis that can be readily applied to both measure and improve the performance of fisheries. In this sense, our book is unique in that it is the only text that systematically explains the economic methods that can be used to analyze fisher behavior. Without hyperbole, we consider the ideas and methods presented in this book are fundamental to securing sustainable and economically viable fisheries.

Our book is a collective effort of four people, but with the support of many others. Two of us are directly involved in fisheries policy in the US and Australia, and all of us have consulted for governments on fisheries management in North and South America, Australia and New Zealand, Europe, Asia, and the Pacific Islands, including providing direct advice to senate and parliamentary committees and cabinet ministers. We also have extensive experience with analysis and management of international fisheries and protected species, such as the Food and Agriculture Organization of the United Nations (FAO) and the WorldFish Center, and international treaty negotiations. In addition, the book benefits from our combined 60 plus years of experience in the analysis of dozens of fisheries from many different countries and regions. All of the examples we use in the book are extensions of the many papers that we have published in fisheries economics. Throughout the text we use real-world cases and data to give you, the reader, insights into the challenges (and also the rewards!) of the economic analysis of fisheries.

The book is structured around six self-contained chapters. The introductory chapter outlines the bioeconomics of fishing and different management approaches to fisheries. Chapter 2 provides a comprehensive review of the sort of data that is required to undertake the economic analysis of commercial fisheries, how this data can be collected, and in what ways the data should be managed and used. Chapter 3 explains how efficiency measures can be obtained for individual vessels and fishing fleets, and how this might be used to improve management. Chapter 4 focuses on the measurement of capacity and capacity utilization in fisheries and what it implies for fisheries management. Chapter 5 describes an index number approach to assessing productivity performance in fisheries and explains how it can be used. Our concluding chapter puts the three approaches (efficiency, capacity and productivity analysis) in context and gives insights about what the economic analysis of fisheries can contribute to the various aspects of management. We believe that, collectively, the six chapters in our book provide both the understanding and the tools required to substantially improve fisheries management.

Acknowledgements

The book is the joint work of all four authors, but we also received help from two others: Stephen Stohs was a collaborator and co-author on Chapter 2, and Catherine J. Morrison Paul provided valuable input and assistance on Chapter 4. Tuong Nhu Che was helpful to us in many aspects of the book while Kevin Fox reviewed and provided useful comments on Chapter 5.

We both acknowledge and thank everyone, especially Stephen and Cathy, and also the commercial fishers, fisheries managers and scientists who have helped us to understand the realities of fishing and the difficulties of management. We alone accept responsibility for any errors of commission and omission.

List of Abbreviations

ABARE	Australian Bureau of Agricultural and Resource Economics
AFMA	Australian Fisheries and Management Authority
BC	British Columbia
CPUE	Catch per unit of effort
CU	Capacity Utilization
DEA	Data Envelopment Analysis
EEZ	Exclusive Economic Zone
EKS	Elteto-Koves and Szuc (method)
FAO	Food and Agricultural Organization of the United Nations
GDP	Gross Domestic Product
ITQ	Individual Transferable Quota
IVQ	Individual Vessel Quota
MCC	Maximum Carrying Capacity
MEY	Maximum Economic Yield
MP	Marginal Product
MSY	Maximum Sustainable Yield
NPF	Northern Prawn Fishery
PV	Present Value
Qty.	Quantity
SCPUE	Stock-adjusted catch per unit of effort
SFA	Stochastic Frontier Analysis
TAC	Total Allowable Catch
TC	Total Cost
TE	Technical Efficiency
TR	Total Revenue

Chapter 1

The Economics of Fishing and Fisheries Economics

Historically, the emphasis was given to the fish. More recently, it has been seen to be necessary to pay more attention to the complex of social, economic, and political factors that drive the behavior of fishermen as individuals and fisheries as systems.
Peter Larkin (1978, p. 57) in Fisheries Management – An Essay for Ecologists.

1.1 Introduction

Many of the world's fisheries are challenged by a combination of overcapacity, overharvesting, habitat damage and poor economic returns. For the first time ever it appears the world's total harvest of fish from wild stocks is in decline because of over-fishing (Hilborn et al. 2003). The challenge is such that over a decade ago the Food and Agricultural Organization of the United Nations noted '... many conservatively-targeted quota management systems have failed, even for proprietary resources of EEZs, ...' (FAO 1993, p. 37).

In an attempt to address these difficulties managers are looking for innovative ways to address the 'tragedy of the commons' whereby individuals operating in their own self interest overexploit a common-pool resource which is open to all (Hardin 1968). At the forefront of 'new' thinking about fisheries is the so-called ecosystem approach (Pikitch et al. 2004) that places a much greater weight on integrating management across fisheries and maintaining healthy ecosystems (habitats, biodiversity, resilience to shocks, etc.).

While an integrated approach to fisheries is helpful, and new approaches are most certainly required to prevent further declines (Pauly et al. 2002), without careful attention by regulators to fisher incentives little will be accomplished. We contend that the biggest barrier to sustaining fisheries is not our ignorance of marine ecosystems, nor is it a lack of goodwill from those involved in fishing. Rather, the stumbling block to better fisheries management is a profound lack of understanding of the behavior of the oceans' greatest predator – the commercial fisher.

This chapter, and indeed the entire book, is about how to understand and how to analyze fisher behavior from an economic perspective. Too frequently managers are perplexed by the lack of support by fishers for necessary cuts in the Total Allowable Catch (TAC). If fishers protest against cuts in the current TAC, even when this imposes significant risks for the sustainability of the stock, a fisheries economist would not presume that fishers are acting irrationally. Rather, she would examine what are the incentives and motivation for fishers to act in such a myopic fashion. For example, in the New England groundfish fisheries many commercial fishers oppose cuts in the TACs despite the fact that lower catches for a while would likely result in higher catches in the future. This is not irrational behavior, but simply fishers responding to incentives. Given that the New England groundfish fisheries are primarily regulated by input controls, especially limits on days at sea, if the currently active fishers 'stinted' or lowered their harvests they would have little assurance that they would be the beneficiaries of such conservation (Hilborn et al. 2005). Indeed, with a large amount of latent effort and vessels not currently active in the fishery, but which would have the necessary licences to enter the fisheries should it become profitable to do so, it makes little sense for existing fishers to support catch reductions that impose real short-term costs for very little expected future benefit.

Our contention is that a much greater attention to fishers, their incentives and how regulations affect fisher behavior will pay handsomely in terms of more sustainable fishery outcomes. Obviously other approaches are also required to promote sustainability, such as the use of marine reserves (Grafton et al. 2005), but a greater focus on fishers will go a long way to support healthy marine ecosystems and improve the economic performance of fisheries.

In this chapter we present a model of the economics of fisheries that describes why many fisheries are overexploited from both a biological and economic perspective. We also explain what we mean by fisheries economics, and briefly describe the principal approaches we will use in this book to measure and analyze fisher behavior and fishery performance.

1.2 An All Too Common Tragedy

The history of fishing is replete with examples of fisheries that have been exploited to commercial extinction (Schiermeier 2002). The basic cause is not the rapaciousness of fishers, but principally arises from the characteristics of fisheries – harvests are rivalrous, fish are fugitive and thus are difficult to 'own' and manage, and fisheries are subject to irreducible uncertainties. The rivalry in fishing comes from the fact that fishers harvest from a resource limited in size such that what one fisher catches today cannot be caught tomorrow by somebody else. This is sometimes called the 'common-pool' problem because each fisher is using a common resource in which the yield, at a given stock size, is more or less fixed by nature.

In a fishery if a skipper decides not to harvest until later, he potentially benefits all other competing skippers. In general, it would be in the interests of all those who fish to agree to restrict their catch to prevent overexploitation. Such a cooperative outcome, however, is difficult to enforce and 'shirking' is likely in large-scale fisheries prosecuted by many vessels. In such fisheries, each vessel has an incentive to 'free ride' once a deal has been struck, by increasing harvest while others are reducing theirs. The problem with fishing is that it is difficult to prevent or exclude others from using the resource. Unlike with land, for example, a fence cannot be placed around fish in the sea and the costs of excluding other users are much higher than with many terrestrial resources.

The absence of one, property rights over the fish, two, effective management of the resource, three, cooperation among harvesters and four, 'free' entry into a fishery by outsiders is commonly called an *open access* resource. In open access, the harvesting costs imposed on others are not taken into account by fishers when they make their decision as to how much fish to catch. These costs include less fish for others to catch, but may also encompass habitat damage, and also congestion costs if vessels all try to catch fish at the same locations at the same time. Such costs are what economists call *negative externalities* and suggest that, in the absence of management, ownership or controls on fishing, there will be too much fishing, and too many fish harvested. In essence, this is what has come to be called the 'tragedy of the commons'.

The policy action proposed to prevent the 'tragedy of the commons' in fisheries has been to restrict access to fishing grounds, and to limit the Total Allowable Catch (TAC) by fishing fleets. The goal has been to prevent over-fishing and ensure sustainable harvesting. Despite a huge expenditure by regulators totalling billions of dollars to determine stock sizes and the appropriate TAC, there have been several spectacular stock collapses. One of the largest and most recent was the northern cod fishery off the coast of Newfoundland and Labrador in the early 1990s, which is still at a tiny fraction of even its depleted size in the 1980s. Stock collapses are the most extreme examples of management mistakes, but they exist in one form or another in most fisheries.

A principal cause of collapses, and other problems in fisheries, is the lack of appropriate incentives and institutions that encourage fishers to behave in a sustainable way. The consequences of poor institutions include a lack of resilience – characterized by a poor ability of fish stocks to bounce back following a downturn or a negative environmental event – and also habitat damage, poor average returns to fishers, conflicts among fishers and undesirable fishing practices.

1.3 An Economic Perspective of Fisheries Management

The traditional approach to 'managing' fisheries has been to place the fish before the fisher. In many fisheries, the number one priority has been to maintain fish stocks, and it was presumed that by controlling fishing effort this goal could be achieved. Regulations that restrict the number of vessels fishing, the gear used by fishers or the time spent harvesting have been implemented in hundreds of fisheries and dozens of countries in the past 40 years. The common assumption with such controls is that, if implemented with sufficient vigour, they prevent further increases in fishing effort and ensure sustainable harvests. By the 1970s it was apparent that input controls were failing to prevent what was called 'effort creep' or the slow, but inexorable, increase in fishing effort and capacity as fishers substitute from regulated to unregulated inputs (Wilen 1979). The 'remedy' prescribed by many has been to impose even greater controls, more rigorously enforced. Unfortunately, although this approach has increased both management and harvesting costs it has often failed to prevent further increases in fishing effort (Townsend 1990).

To ensure the desired TAC is not exceeded, regulators frequently resort to shortening the fishing season because the fishing fleet is able, with effort 'creep', to catch more fish per day with each passing year. In response to an ever shortening fishing season, and the need to harvest a certain quantity of fish to cover their fixed costs, especially debt payments, harvesters frequently devise more elaborate ways to substitute from regulated to unregulated inputs and gear. For example, if regulators impose a maximum vessel length fishers respond by building vessels with a much larger width. If limits are placed on the type and size of gear that can be used, harvesters increase their engine size to reduce the time to reach fishing grounds. Whatever the regulation or control, fishers in almost all cases appear to be one step ahead in terms of substituting to alternative inputs to catch their share of the limited catch – a case of 'removing the net long after it has been set'.

As we will show, the economic perspective is *not* that fisheries should be left unregulated, but that the regulations must explicitly consider the incentives of fishers. The 'tragedy of the commons' and the failures associated with open access make it all too clear that a 'laissez-faire' approach to fisheries does not work. Instead, the costs that fishers impose on others from their harvesting needs to be made perfectly clear and, ultimately, some of these costs will be borne by consumers in terms of higher prices. For fishers to behave sustainably, they need long-term and secure rights that explicitly account for interactions across stocks, and also a participatory mandate in management. Improved tenure, forcing harvesters to pay the costs of fisheries adjustments and providing fishers with decision-making responsibilities will help to align incentives with sustainability goals and improve fishery management outcomes.

An economic perspective of fisheries management is that marine resources should be managed sustainably, but also in a way that they contribute to and provide net benefits for the nation as a whole. Indeed, we argue that sustainable and economically profitable fisheries are complementary. A level of harvest that maximizes the sustainable returns from fishing is often at a stock size that is *greater* than that which would maximize the overall yield from a fishery. Moreover, if there are other costs associated with fishing such as habitat damage, or biodiversity and environmental losses such as from by-catch of sea birds, dolphins or turtles, the economic optimum level of harvest that accounts for these costs would be even less, and the desirable fish stock even larger. In other words, a fishery that is economically viable in the long run is also likely to be an ecologically sustainable fishery. As surprising as it might seem to some, economics is the friend of conservation!

1.4 Economics of Fishing

Key to understanding the economics of fishing is the concept of Maximum Economic Yield (MEY) which provides a benchmark to compare current with potential economic performance in fisheries. MEY coincides with the level of harvest or effort that maximizes the sustainable net returns from fishing. A MEY harvest is desirable because it is the catch level that enables society to do the best it can with what nature has provided. By generating the highest possible economic surplus, commonly called the resource rent, the funds can be spent on those goods and services that contribute to overall welfare. For example, in Iceland many of the fisheries generate a resource rent (Arnason 1995) some of which is taxed and used to pay for schools and hospitals. If there were no economic surplus, because of poor fisheries management, the wellbeing of all Icelanders would be diminished. By contrast, the groundfish fisheries of Newfoundland and Labrador off the east coast of Canada have been badly managed. Instead of generating a surplus that could be used to improve the wellbeing of New Foundlanders, the other provinces of Canada have provided billions of dollars in transfer payments in income support to fishers and their families (Hannesson 1996).

Attempts to extend resource use and particularly employment well beyond MEY are common, and often disastrous. Experience in Canada's Atlantic fisheries provides a striking example. Subsidies provided by the Canadian government – with a specific mandate to maximize employment levels in the industry – greatly extended the amount of fishing effort. Indeed, three decades ago it was '…estimated that Canada's commercial catch in 1970 could be harvested by 40 per cent of the boats, half as much gear and half the number of fishers' (Auditor General of Canada 1997, pp. 4/15). This is wasteful in itself, but dwindling stocks and the eventual collapse of the Atlantic fisheries – in large part due to over-fishing – further increased the government's burden to maintain incomes. In 1990, for example, and before the collapse of the groundfish stocks, self-employed fishers received $1.60 in unemployment insurance benefits for every dollar earned in the fishery. The subsequent collapse resulted in 'adjustment programs' that have cost Canadian tax payers billions of dollars (Auditor General of Canada 1997).

The management structure, stock level and nature and extent of fishing effort that generates MEY depends on a combination of biological and economic factors. For the resource rent to be maximized, it must also be the case that the fishing fleet uses its fishing inputs in combinations that minimize the costs of harvest at the MEY catch level. In other words, at MEY fishers are at maximum economic efficiency and there is no overcapitalization of vessels or gear.

To understand what the MEY target is, we provide a brief review of a population model that is sometimes used in fisheries – the surplus production model.[1] The implications of the model are illustrated in Figure 1.1. It shows the yield or net additions to the stock of fish on the vertical axis and the stock of fish on the horizontal

---

1   See Gulland (chapter 3, 1983) for a review of such models.

axis, also measured in the same units (tons of fish). We also assume, for the moment, there is no uncertainty about the state of nature.

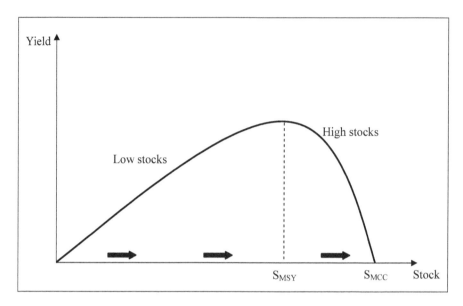

Figure 1.1 Surplus-production Model

The curved line in Figure 1.1 shows the growth in the stock of fish, or yield, for every possible stock size. It illustrates what is normally referred to as density dependent growth. At low stocks, recruitment is small because there are relatively few fish available to reproduce. Recruitment rises as the stock of fish increases, and then falls as the stock of fish begins to 'crowd' the environment and reaches a maximum limit at its carrying capacity. The stock at maximum carrying capacity (S_{MCC}) thus defines the maximum number of fish (or weight of fish depending on the units we are using) that the environment will support. With no fishing, the stock of fish will naturally increase – represented by the arrows moving in the right-hand direction to this point. A sustainable harvest occurs when harvest matches yield, or catch is just sufficient to capture new additions to the stock of fish, at any given stock level. In this sense, each point on the yield curve in Figure 1.1 represents a potential sustainable harvest; with the stock at maximum sustainable yield (S_{MSY}) generating the largest potential catch.

To translate Figure 1.1 into economic terms, we assume that the price of fish is given – as would be the case for a fishing industry that is competitive and faces a world price for its catch – and, for convenience, we set it equal to one dollar. In this case the yield curve, representing sustainable harvest levels, is also an exact measure of the total revenue from each sustainable catch. We also find it convenient to measure fishing effort (such as nominal days fished or trawl hours) on the horizontal axis, rather than stock size. We note that increases in effort result in a fall in stock such that the two variables generally move in *opposite* directions. Accordingly, Figure 1.2 measures Total Revenue (TR), which is a function of fishing effort, in dollars (or Euros or pounds) on the vertical axis and fishing effort on the horizontal axis.

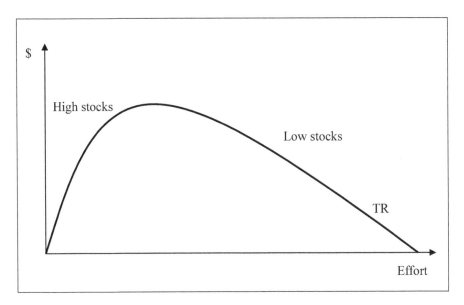

Figure 1.2 The Relationship between Total Revenue and Effort

The stock-yield diagram from Figure 1.1 has thus been flipped 180 degrees to generate Figure 1.2. The origin in Figure 1.2 now represents the fish stock at the maximum carrying capacity because it coincides with zero fishing effort, while the intercept with the largest amount of effort corresponds to a zero stock of fish. By contrast to Figure 1.1, a stock of fish that is plentiful, or with 'high stocks', now occurs on the left-hand side of the diagram in Figure 1.2 and corresponds to low levels of fishing effort, while a low level of fish stocks occurs on the right-hand side of Figure 1.2 is consistent with high levels of fishing effort.

Nothing yet has been said about the costs of fishing. Again, to keep things simple, assume that all fishing vessels are identical and that the Total Cost (TC) of fishing – including the cost of fuel, crew, bait, gear, etc. and also the opportunity cost of using vessel capital and all other inputs that accounts for the 'normal rate of return' on investment – is proportional to the amount of effort applied. Under this assumption we can combine TR and TC in one diagram, as in Figure 1.3. A key feature of Figure 1.3 is point B, called the bionomic equilibrium, which corresponds to the level of fishing effort where total fishing cost equals total revenue. This bionomic equilibrium is the point where economic profit (allowing for the opportunity cost of investment and thus is distinct from accounting profit) is zero. If all the economic profit is attributable to the fishery itself, rather than any unique expertise of any individual or group of fishers, this economic surplus is called the resource rent.

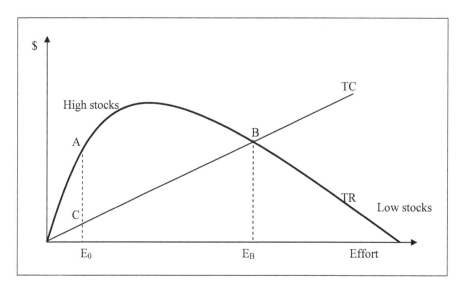

Figure 1.3 Bionomic Equilibrium

Why is point B an equilibrium, or resting point, for the fishery? First, it represents a sustainable harvest. Second, points to the right of effort levels at B will necessarily imply that total costs are larger than total revenues, or that profits are negative. This must imply that it would be better for firms to employ their capital in their next best alternative. In any case, with negative profits, fishers will eventually go bankrupt and leave the fishery until point B is again obtained. In the case where vessels differ, those that are the least efficient, or have the highest cost of fishing, will be the first to exit.

Points to the left of B are more interesting, and illustrate the 'tragedy of the commons' that is associated with every bionomic equilibrium. Begin, for example, at an initial effort level E_0, where profits are positive and are measured by the distance A to C. Profits are large in this case because stocks are 'thick' and the cost of fishing is relatively low. Costs are low at this low level of effort because 'thicker' stocks mean less time is spent fishing and higher catches per unit of effort (each cast of the net, so to speak, catches more fish) lowers the cost per unit of harvest. However, in an unregulated or open access fishery, the existence of positive economic profits – over and above the average rate of return that can be obtained elsewhere – induces new fishing vessels to enter the industry, and those vessels already in the fishery to expand effort and capture the extra profit. As long as profits are positive, this process of effort creep will continue up and until point B, where there is no further incentive to expand effort. When all vessels act in this way, the stock of fish falls over time and the per-unit cost of fishing rises until all economic profits are dissipated.

If any skipper decides to limit fishing effort and conserve stocks, while others do not, that fisher will be relatively worse off. All fishers, acting in their own self-interest, are induced to fish more, but because those harvesters that increase effort do not take into account the effect of their fishing activity on other vessels in the fishery – including the increased cost of harvest as a result of stock depletion – eventually all vessels are worse off. Indeed, in this sense, point B is undesirable for at least two reasons: first, because profits are zero and the cost of fishing is needlessly high and second, as drawn – and this does not necessarily have to be the case – it would have been possible to obtain the same catch with less effort, lower costs and larger stocks at point A.

The case of a bionomic equilibrium makes it clear how profits can be maximized in a fishery, or how to find MEY. In Figure 1.4 this occurs at the effort level E^* that corresponds to a value of catch $\$_R$ that creates the largest difference between the total revenue and total cost of fishing. This level of effort maximizes economic profits or the resource rent per unit of effort, and is given by the difference between $\$_R$ and $\$_C$, or R^*.

MEY at point A in Figure 1.4 implies that not only are profits maximized at A, but the value of harvest (both yield in physical terms and the value of catch in terms of revenues) has also increased compared to the bionomic equilibrium at point B. The reason that profits are now larger at point A is not only because TR is higher, but given that stocks of fish are larger and the amount of 'days' spent fishing is smaller, the cost of fishing is also less, as shown by the move from point B to C. Such a 'win-win' outcome (higher TR and lower TC with a fall in catch) does not always occur with a move from the bionomic equilibrium to MEY. In many fisheries the cost of fishing may already be sufficiently high (simply rotate the TC curve closer to MEY, implying a fall in effort at MEY) such that a move from a bionomic equilibrium to MEY will require a fall in both the harvest and revenues, but not economic profits.

By redrawing Figure 1.4 on page 10 we can show that MEY is affected by changes in the price of fish and the costs of fishing. An increase in the price of fish, for example, results in a shift upward of the TR curve at all effort levels, leaving the intercepts unchanged. For a given cost curve, the point of MEY moves closer to MSY, but in this figure never beyond MSY so long as the cost of fishing increases with effort. In other words, the more valuable landed fish are, the more it pays to work the fishery harder, and thus decrease the equilibrium stock of fish. By contrast, an increase in costs, or a rotation upwards and to the left of the existing TC curve, the new MEY point moves to the left of the previous MEY and lowers optimal fishing effort because with a more costly harvest, it pays to have larger stocks from which to catch. By contrast, technological change that reduces harvesting costs downwards to the right the new MEY point moves to the right of the previous MEY and increases the optimum level of fishing effort. In sum, a fall in the price of fish, or an increase in harvesting costs, implies a lower harvest with less fishing effort and a larger stock size in order to maximize economic profits.

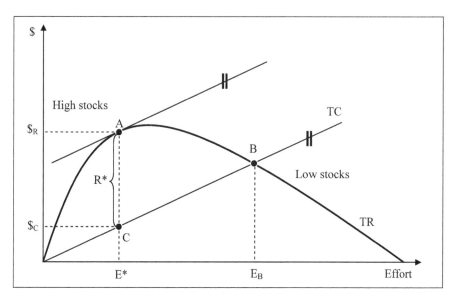

Figure 1.4 Maximum Economic Yield (MEY)

The discussion of MEY underscores the undesirability of MSY (and other biological indicators) as a target, at least as long as having an economically viable fishing industry is an objective. Pursuing MSY alone can result in zero, or even negative, profits at that target level. For instance, it is possible that at a sufficiently high cost of fishing, the fishing effort at MSY can result in the total cost greater than the total revenue. In the long run, no fishing industry will move beyond the bionomic equilibrium given by point B because profits would turn negative (although cases where average costs exceed average revenues for a period of time are not uncommon in poorly regulated fisheries). However, if a biological target (such as MSY) is consistent with negative profits this cannot be a good management goal. Indeed, in the case of a bionomic equilibrium to the left of MSY, where effort is measured on the horizontal axis, a regulatory environment that attempts to target and enforce MSY will simply result in a replication of the bionomic equilibrium. This is clearly undesirable, especially when we consider the considerable resources (financial and scientific) required to estimate and implement an MSY.

1.5 Beyond Maximum Economic Yield

There are three important qualifications to the economics of fishing and the MEY target. So far we have assumed, one, a zero rate of discount, two, the cost of fishing depends on stock size in a simple linear fashion and, three, fishing costs rise proportionately with effort. The discount rate is the interest rate at which future income or catches are valued today. A plausible case can be made for a zero discount rate in a well-managed fishery, but fishers and regulators certainly operate as if there is a positive interest rate, and act as if a harvest some time in the distant future is worth less than a harvest today. Given a positive discount rate, a modified version of MEY is appropriate and will tend to move optimal effort and catch closer to the bionomic equilibrium. In other words, if current catches are valued more highly than future harvest it pays to work the fishery harder today, with smaller equilibrium stocks of fish. It is even possible that if the discount rate is high enough, MEY will correspond to stocks that are smaller than those associated with MSY.

In general terms, it is not hard to show that if the discount rate becomes very large indeed, MEY will correspond to a bionomic equilibrium (Clark 1990). This is because it pays to draw down the stock today if the loss of future net returns are very heavily discounted. For instance, in an open access fishery each individual fisher completely discounts the future, at least in terms of rate of harvest because if fish are left in the sea they will be caught by someone else. A discount rate in between zero and the rate that makes the bionomic equilibrium an optimum will place MEY somewhere between E^* and E_B in Figure 1.4. In practice, for most fisheries that are productive – with reasonably large intrinsic rates of biological growth – and with discount rates that reflect normal rates of return (say 5 per cent or less), it is often, but not always the case, that this modified or dynamic MEY will occur to the left of MSY in Figure 1.4. In other words, for many (but not all) fisheries with commonly used discount rates, MEY will normally be more 'conservationist' and generate stock sizes that are *larger* than those associated with MSY, or a comparable biological target. This implies that maximizing the economic viability of fisheries is often compatible with maintaining the ecological sustainability of fisheries. Moreover, a 'thicker' or larger fish stock associated with a MEY target can help protect fisheries and increase resilience in the face of unforeseen, or negative, stochastic environmental shocks that may diminish the fish population.

The conservationist implications of the MEY target point are strengthened if relevant cost considerations are also taken into account. The implication of a cost of fishing that *increases* with stock depletion, at an increasing rate – what economists refer to as convex cost functions in terms of stock – probably characterize most fishing activity. Given such costs it is economically desirable to have a catch level and effort even further to the left of the bionomic equilibrium. If it is more costly to fish as stock decreases, and if this cost increases at an increasing rate, it pays to have even larger stock sizes than that depicted at MEY in Figure 1.4. This will partly offset (and in some cases even more than offset) the effect of a positive discount rate.

Another important caveat is that, so far, we have implicitly assumed a single species fishery. Multi-species fisheries create complications in a number of ways. If species interact biologically this calls for relatively complicated approaches, such as predator-prey models, where the notion of sustainability itself becomes difficult to define. If the interactions mostly occur 'above the water', so to speak, or in terms of the profitability of vessels, the bioeconomic model must also account for differing prices across species, the value of target versus bycatch species, effort split across target species, and the likelihood that the cost of fishing, and specific cost functions, vary across individual species. It is possible to account for all of these factors, but determining the value of MEY for each species becomes much more difficult.

Finally, our analysis assumes that the population biology and all of the relevant economic functions and parameters were drawn from a deterministic setting, or one with no uncertainty about the state of nature or the economics of the fishery. In reality, this is never true. Important sources of uncertainty include a lack of complete biological data to calculate the stock-recruitment relationship (the yield curve in Figure 1.1) and inability to accurately measure the actual catch or effort of fishers and also the size of the current fish stock. In some cases, natural variability in stocks may make it all but impossible to estimate a yield curve, and thus the relationship between total revenue and effort. Another source of uncertainty is the price of fish and the precise cost of fishing. These must be forecast, and forecast errors are common. If these errors are systematic, then efficiency gains from targeting MEY will diminish. After taking uncertainty into account, it is not unreasonable to approach an estimated MEY target in a slow way, with adaptive management responses to changes in prices, costs and the underlying biology of the fishery.

Confronting uncertainty in fisheries also calls for a different approach to how fisheries are managed. It requires that regulators consider the questions: how, when and where fish are caught, rather than simply focusing on the total harvest (Wilson et al. 1994). It also demands an adaptive approach to management and the use of marine reserves, to promote resilience in the face of negative environmental shocks (Grafton and Kompas 2005). Moreover, it requires an appreciation that a mix of strategies is needed to minimize the risk of management failure, irrespective of the underlying stock or economic dynamics. A mixed strategy approach that includes, for example, controls on total catch, age and size and sex of fish caught, season length, closed exploitation areas and participation controls that explicitly account for fisher incentives, can provide managers with options given uncertainty (Grafton and Silva-Echenique 1997).

Despite our caveats, pursuing an MEY target in fisheries is important. It not only helps protect the fish population, by ensuring that stock levels in many fisheries are larger than those associated with the traditional MSY target, but also ensures that valuable inputs (such as fuel and labor) are allocated efficiently, and in a manner that maximizes profits. Management regimes that attempt to extend the amount of resources devoted to the fishery beyond MEY only generate a system with excess boat capital and lower returns from fishing. In many cases – especially those where input restrictions fail to prevent effort creep – the fishery simply 'winds down' to a state where total fishing days are severely limited and asset values and profits are low. In such cases, the fishing industry often calls for government assistance or some sort of vessel buyback scheme to restore profitability (Holland et al. 1999). Unfortunately, the history of buyback schemes is that even if the regulator is successful at removing excess fishing effort and increasing profits, the increased returns with input controls encourage further effort creep thereby eroding the initial gains (Grafton and Nelson 2004).

1.6 Bioeconomic Modeling and the MEY Target

To this point we have assumed that the management authority can simply calculate the desirable MEY. If a fishery is *not* in equilibrium or in a steady state such that we know the exact yield for given fish stock size, calculating the MEY can be both an elaborate and time-consuming process that involves both biological and economic modeling.

To illustrate how an MEY target can be calculated we provide two approaches. In the first and most preferred approach we use data from the Northern Prawn Fishery (NPF) located off Australia's northern coast to calculate MEY, following Kompas and Che (2004). As this type of bioeconomic modeling is data intensive and requires some expertise, we also describe another, simpler approach. However, the second approach has the very considerable drawback that it assumes that the fishery is in a steady state or equilibrium such that the available fishing data provide an accurate representation of the stock-sustainable yield relationship.

Dynamic Bioeconomic Model of the Northern Prawn Fishery

Although there are more than fifty species of prawn that inhabit Australia's tropical northern coastline, only about nine species of prawns are caught. Commercial prawn species have a life span of one or two years while juvenile prawns live in coastal and estuarine areas, in beds of seagrass or mangrove-lined creeks. Three species (the white banana prawn *Fenneropenaeus merguiensis*, the brown tiger prawn *Penaeus esculentus*, and the grooved tiger prawn *P. semisulcatus*) account for almost 80 per cent of the total annual landed catch weight from the fishery. Endeavour prawns (*Metapenaeus endeavouri* and *Metapenaeus ensis*) and the red-legged banana prawns (*F. indicus*) form most of the remainder of the catch. Other commercial catch includes the giant tiger prawn (*P. monodon*), western king prawn (*Melicertus latisulcatus*) and the red spot king prawn (*Melicertus longistylus*).

To date, a series of input controls have been used in the management of the fishery. These include limited entry and gear restrictions (through the issuing of Statutory Fishing Rights), a system of spatial and temporal closures, and bycatch restrictions. The NPF has two open seasons when fishing is permitted. The seasons have varied in the past, but are generally April to May/early June and August to November. The closure of the fishery generally coincides with the spawning and recruitment phases of prawns to ensure individuals are at an acceptable size for harvesting. Shortening of the season has also been used as a method to reduce effort in the fishery. Other regulations include the permanent closure of seagrass beds (and other sensitive habitats) and seasonal closures of juvenile prawn stock habitat (AFMA 2003).

In its relatively short history the fishery has experienced a significant variation in catch. With this, falling prawn prices and profitability in the 1980s led to a drastic restructuring of the fleet in the late 1980s and early 1990s. In addition, the fleet structure changed gradually throughout the 1970s, with a move away from wooden trawlers with brine tanks and ice boxes toward larger, purpose-built, steel freezer trawlers with high catch and carrying capacities.

(1) Biological Model
To estimate MEY for the NPF, a spawning stock-recruitment relationship is modelled as follows,

$$R_t = \alpha_1 \times \hat{S}_{t-1} e^{\beta_1 \hat{S}_{t-1} \xi} \tag{1.1}$$

where R_t is the total number of recruits produced in year t and $\hat{S}_{t-1}$ is the spawning stock of the previous year (estimated as the number of prawns). The parameters α_1 and β_1 need to be estimated to calculate the stock-recruitment relationship, but we would expect $\alpha_1 > 0$ and $\beta_1 > 0$. The measure ξ reflects uncertainty or the stochastic behavior of the spawning stock-recruitment relationship, usually drawn from a probability distribution.

Much like in Figure 1.1., the stock-recruitment relationship has a concave shape reflecting density dependence. In other words, at small levels of the spawning stock, a larger stock increases recruitment, but beyond a certain stock threshold where recruitment is at its maximum further increases in the spawning stock lowers recruitment.

The previous year's spawning stock is assumed to be a fixed proportion (denoted by γ) of the total female stock where female prawns are assumed to constitute half of the total stock of prawns (S_{t-1}), i.e.,

$$\hat{S}_{t-1} = \gamma(\frac{1}{2} \times S_{t-1}) \tag{1.2}$$

Following Wang and Die (1996), we also specify that the current spawning stock $\hat{S}_t$ is a function of annual recruitment R_t and fishing mortality, defined as

$$\hat{S}_t = \alpha_2 \times R_t \times e^{-\beta_2 (F_t + m)} \tag{1.3}$$

where F_t is the fishing mortality at year t, m is the annual natural mortality rate and α_2 is a parameter. In turn, fishing mortality at year t is defined by

$$F_t = q \times E_t = q \times B_t \times N_t \tag{1.4}$$

where q is the 'catchability coefficient' and E_t is fishing effort at year t. Fishing effort is defined as the total 'standard' boat days in the fishery which is the product of total 'standard' boats (B_t) and nominal fishing days in the season (N_t). In this model, one unit of fishing effort is defined as the daily effort of a 'standard' boat and is used to avoid the problem of equating boat day units between large and small vessels.

If we assume that fishing effort changes over time due to technical change then (1.4) can be rewritten as,

$$F_t = q \times E_t = q \times TEC_t \times B_t \times N_t \tag{1.5}$$

where TEC_t is the variable that measures the change in technology at year t. Finally, we assume that the annual catch h_t in tonnes depends on recruitment and natural and fishing mortality. Given that fishing mortality itself depends on fishing effort, we can write the level of harvest as a function of fishing effort, i.e., $h_t = h(E_t)$

(2) Economic Model
Annual total revenue (TR_t) of the fishery is defined as the product of annual fish harvest and the annual (average) price of prawns, i.e.,

$$TR_t = p_h h(E_t) \tag{1.6}$$

where p_h is the unit price of prawns and is inversely related to the total catch. Annual total cost is assumed to be the sum of labor, material, capital and other costs. Labor costs generally include a share of total fish revenue and packaging and gear maintenance expenditures. Capital costs[2] and other costs (of which fuel is a major component) are assumed to depend on fishing effort so that total costs can be expressed as follows,

$$TC_t = \alpha + c_L h(E_t) p_h + c_M h(E_t) p_h + c_K E_t + c_O E_t \tag{1.7}$$

---

2   Capital costs are defined by the cost of capital calculated as a sum of depreciation cost and the annual opportunity cost of boat capital value.

The parameters c_L and c_M are the costs shares for labor and fishing materials per each Australian dollar of output, and c_K and c_O are parameters that represent the average capital cost and other costs per unit of effort. The average capital cost of a unit of effort (c_K) is estimated by dividing total capital costs by total effort. Average other costs (c_O) per unit of effort is estimated by dividing total other costs by total fishing effort. Fixed costs are measured by the parameter α.

Annual fishery profit (Π_t) is annual total revenue less annual total cost, i.e., $TR_t - TC_t$. To determine MEY, we maximize expected profits over time, which involves incorporating uncertainty with respect to the stock-recruitment relationship and catch per unit of effort into our optimization. We also discount future costs and benefits which involves multiplying the profit in each period by a *discount factor* defined as follows,

$$\frac{1}{(1+\delta)^t} \qquad (1.8)$$

where δ is called the discount rate and $(1+\delta)$ is raised to the power of t where t is the number of periods into the future from the present. The larger is the discount rate, the lower will be its value in present value terms (or in current dollars), all else equal. For example, if we wished to discount expected profits four periods into the future with a discount rate of 3 per cent $(\delta = 0.03)$ we would multiply the future expected profits by the appropriate discount factor to obtain a *present value* (PV) of expected profits,[3] i.e.,

$$\text{PV of Expected Profits (period 4)} = \frac{1}{(1+0.03)^4} \times \text{Expected Profits (period 4)}$$

Under existing input control regulations the fishery authority targets overall effort levels through a combination of input restrictions, limits on technology and limitations on days fished. Assuming that effort levels are observable and enforceable, the problem is to solve (1.9) for E_t, or the number of standard boat-days in each period, over the optimization period given by $t = 1, 2, ..., T$ i.e.,

$$\operatorname*{Max}_{E_t} \sum_{t=1}^{T} \widehat{\Pi}_t = \sum_{i=1}^{T} \frac{1}{(1+\delta)^t}\left[p_h h(E_t) - (\alpha + c_L h(E_t)p_h + c_M h(E_t)p_h + c_K E_t + c_O E_t)\right] \quad (1.9)$$

---

3   See Grafton et al. (2004, p. 111) for further details on discounting in fisheries.

where $\widehat{\Pi}_t$ is the present value of annual economic profit in the fishery at year t. The optimal solution is a set of values for E_t where $t = 1, 2, ..., T$ that shift the stock to a steady state value (subject to the stochastic behaviour incorporated in the model) that maximizes the sum of the present value of economic profits from fishing prawns.

(3) Bioeconomic Modeling Results
Using parameter estimates obtained from the NPF and economic data provided by the Australian Bureau of Agricultural and Resource Economics (ABARE) several versions of the model given by (1.9) can be solved. A base case model, assuming no uncertainty in effort or catch, is provided in Table 1.1. The time horizon for the optimization model is 50 years ($T = 50$) which is long enough to ensure that optimal results are sufficiently close to their steady state values, and the assumed discount rate is 3 per cent.

Table 1.1 on page 18 shows two model results: mean values for harvest and days fished at steady state and, because there is a period of convergence to the steady state, average values per year. Actual values (on average) in the NPF are also shown. The results indicate a substantial reduction in the number of total boat days is required to maximize the economic surplus from the fishery. Optimal boats days are 6,460 compared to the current value in the NPF of 8,455. Given that the total number of vessels in the fishery is fixed, the chosen instrument to achieve this target is assumed to be the number of days fished. The problem with this instrument, as with all input controls, is that it fails to change the incentives faced by fishers. As a result, if it is profitable to do so, fishers will substitute to unregulated inputs which results in effort creep and will eventually require the regulator to further lower the length of the fishing season to try and achieve the MEY target. A far better approach is to give fishers the appropriate incentives to minimize their costs of production while ensuring the total catch of prawns is at a level that maximizes the economic yield.

Finally, it is important to note in Table 1.1 that steady state values of harvest and days fished depend on the price of fish and the cost of fishing. Thus the MEY target must be flexible and should adjust to changes in the price of fish and the cost of fishing.

Table 1.1 Optimal Solution of Base-case Model for the Northen Prawn Fishery (Discount Rate is 3 per cent)

	Unit	Optimal Target
Mean values at steady state		
Annual harvest	*Tonnes*	1,680
Number of boats in a year	*Boats*	95
Fishing days per boat per year	*Days*	68
Total boat days per year	*boat-days*	6,460
Average values per year		
Annual harvest	*Tonnes*	1,560
Number of boats	*Boats*	95
Fishing days per boat	*Days*	63
Total boat days	*boat-days*	5,985
Actual (average) values in the NPF		
Number of boats	*Boats*	95
Fishing days per boat	*Days*	89
Total boat days	*boat-days*	8,455

An Equilibrium Approach to Estimating MEY

The so-called equilibrium approach uses a surplus-production model of the type illustrated by Figure 1.1. A general specification of such a model is the General Production Model (Pella and Tomlinson 1969) that relates growth in the fish stock, usually measured in terms of biomass or weight of fish, to a maximum carrying capacity (S_{MCC}), the current stock level (S), an intrinsic growth rate (r) that determines the speed at which the population returns to S_{MCC} if it is below this level, and a 'skewness' parameter (m) which determines how flat the 'tail' is beyond the level of fishing effort that maximizes the sustainable yield. Normally, the growth in the fish stock is defined at an instantaneous point in time, but for practical applications it needs to be applied in terms of a fixed time interval, such as a fishing season, or a year. In this representation the general production model defines the change in the fish stock or biomass over the fixed interval (such as a year), given by ΔS, as a function of the average stock size or biomass over the same interval and defined by $\overline{S}$, i.e.,

$$\Delta S = rS_{MCC}\overline{S} - r\overline{S}^{m} \qquad (1.10)$$

If $m = 2$ then the yield stock-relationship or curve, and also the yield-effort relationship, are symmetric such that the stock level that maximizes the sustainable yield is $\frac{S_{MCC}}{2}$. If $m < 2$ then the yield–stock curve is similar to that portrayed in Figure 1.1 and the yield-effort relationship resembles what is portrayed in Figure 1.2, i.e., the curve shows a long 'tail' beyond the level of effort that maximizes the sustainable yield. If $m > 2$ then the yield-effort curve drops off more steeply to the right beyond the effort level that maximizes the sustainable yield.

If the fish stock is in a steady state, or equilibrium, then the change in the fish stock will exactly equal the level of harvest such that the stock or biomass does *not* change over the period, i.e.,

$$h = rS_{MCC}\overline{S} - r\overline{S}^m \qquad (1.11)$$

Thus equation (1.11) represents an equilibrium relationship between average stock size over a given interval of time and the total harvest over the same period. However, for most fisheries the assumption of an equilibrium relationship is inappropriate because there exists substantial year to year changes in recruitment and natural mortality due to environmental factors, such as changes in ocean currents and temperatures, which are independent of the level of harvesting or the overall stock size. If the stock size did, in fact, change over the period then (1.11) would misrepresent the harvest-stock relationship. For instance, if the fish stock is declining, but we assume that it is in equilibrium, the approach will tend to overestimate the optimal size of the stock and optimum level of fishing effort. This can, and has, contributed to overexploitation of some fish stocks as equilibrium models have given some fishery managers a false sense of security that the regulated level of fishing effort is appropriate, and the target stock size is sustainable.

To determine MEY we also need to specify a harvest equation. The simplest specification we can use assumes that the total catch is proportional to both the level of fishing effort and the average stock or biomass over the specified time,[4] i.e.,

$$h = qE\overline{S} \qquad (1.12)$$

where q is a catchability coefficient that will vary with changes in technology and the composition of the fishing fleet and E is amount of fishing effort (in days fished or some other measure). This harvest specification implies that the catch per unit of

---

4   See Morey (1986) for a detailed discussion on the appropriate harvest functions for fishing.

effort (CPUE) in the fishery is proportional to the average level of the fish stock, i.e.,

$$CPUE = \frac{h}{E} = q\overline{S} \tag{1.13}$$

In fisheries where independent measures of abundance of the stock are lacking, equation (1.13) is sometimes used to provide an estimate of the average stock size as data on total harvest and some measure of level of effort (such as number of vessels × days at sea) are more readily available.

We can also write fishing effort as a function of the average stock size and the total harvest, i.e.,

$$E = \frac{h}{q\overline{S}} \tag{1.14}$$

If we set the right hand side of (1.11), the change in the stock, equal to the right hand side of (1.12), the specified harvest function, we can solve out for the level of effort that would leave the average fish stock unchanged at $\overline{S}$, i.e.,

$$E = \frac{r}{q} S_{\text{MCC}} - \frac{r}{q} \overline{S}^{m-1} \tag{1.15}$$

Provided that we have data on the size of the stock or biomass, and the harvest and fishing effort over a sufficient number of periods, we can estimate a modified form of equation (1.10) by including harvest as an explanatory variable for the change in stock size, and also estimate equation (1.13), to obtain estimates of the parameters r, q and m, and the maximum carrying capacity, S_{MCC}.[5]

Given parameter estimates for r, m, q and S_{MCC}, we can solve for the optimal level of stock (or equivalently the optimal level of effort) that maximizes the economic profit in the fishery. This requires us to find the stock size (or fishing effort) that maximizes equation (1.16),

$$\text{Max } \Pi = ph - cE \tag{1.16}$$

---

5   Details on how this might be done are provided in Hilborn and Walters (1992, pp. 305-14).

To solve directly for the optimal fish stock size that corresponds to MEY we substitute the expression for *h* from (1.11) and the expression for *E* from (1.15) into (1.16) to find the average stock size ($\bar{S}$) that maximizes the economic profit in the fishery. If we know the optimal stock size, we can calculate the optimal level of effort using the expression given by (1.15), and the optimal level of harvest using (1.11).

Apart from the very serious potential bias from assuming an equilibrium when the stock is in decline, the other major difficulty with the equilibrium approach to estimating MEY is that it assumes that the discount rate is zero. Moreover, unlike the dynamic approach presented for the Northern Prawn Fishery, it does not give any guidance as to the transition of the level of fishing effort for all future periods until we get to MEY.

1.7 An Economic Guide to Fisheries Management

To implement an actively adaptive approach to fisheries management, regulators need clearly defined goals, strategies and tactics to achieve these objectives, and a process to evaluate performance to adapt to change. Fisheries economics can help in all of these management phases. For example, economic analysis allows managers to, one, generate performance criteria to quantify management goals, two, employ models to compare management strategies and tactics and three, use methods to calculate performance criteria and adjust policies and regulations to better achieve targets.

It is worth emphasizing that economics is valuable even if fisheries regulators choose not to target harvest at the MEY, but the analysis becomes even more valuable when managers have explicit economic goals. In other words, if managers choose MSY or some other biological target, the economic analysis of fisheries will still help managers to achieve these biological objectives more effectively. For instance, whatever the biological harvest target, economics analysis can help set incentives such that fishers no longer find it profitable to 'race to fish' that generates effort creep and overcapacity (Grafton et al. 1996).

Fisheries economics covers many topics from bioeconomic modeling to game theory to the economics of marine reserves. No single book can hope to cover all these topics, even in a cursory way. Our objective is less ambitious, but nevertheless a daunting one, to provide the tools of analysis to enable fisheries managers and scientists to understand the economics of fishing to better manage fisheries. To achieve this goal, we focus on four key areas: the collection of fisheries data necessary to undertake economic analysis (chapter 2); and the measurement and application of efficiency (chapter 3), capacity and capacity utilization (chapter 4), and productivity (chapter 5) in fisheries.

Efficiency, capacity and productivity analysis are the pillars of the economics approach to fisheries along with bioeconomic modeling. The three approaches complement each other and can be used to measure both economic and biological performance, to evaluate existing strategies and tactics, and to give insights as to how to improve fishery outcomes. We present these methods using real-world examples to show what insights economic analysis can provide. Collectively, the methods and the examples in our book represent an economic guide to fisheries management that will help researchers, managers and scientists to achieve more profitable, but also more sustainable, fisheries.

References

AFMA (Australian Fisheries and Management Authority), *Northern Prawn Fishery: Strategic Assessment* (Canberra, 2003).

Arnason, R., *The Icelandic Fisheries: Evolution and Management of a Fishing Industry* (Osney Mead, UK: Fishing News Books, 1995).

Auditor General of Canada, *Atlantic Groundfish Fisheries* (Report of the Auditor General of Canada to the House of Commons, Minister of Public Works and Government Services, Ottawa, 1997).

Clark, C.W., *Mathematical Bioeconomics: the Optimal Management of Renewable Resources* (New York: John Wiley & Sons, 1990).

FAO (Food and Agricultural Organization of the United Nations), *Reference Points for Fishery Mismanagement: their Potential Application to Straddling and Highly Migratory Resources* (FAO Fisheries Circular, no. 864, Rome, 1993).

Grafton, R.Q., Adamowicz, W., Dupont, D., Nelson, H., Hill, R.J., and Renzetti, S., *The Economics of the Environment and Natural Resources* (Malden, MA: Blackwell Publishing, 2004).

Grafton, R.Q. and Kompas, T., 'Uncertainty and the Active Adaptive Management of Marine Reserves', *Marine Policy*, 29 (2005): 471-9.

Grafton, R.Q., Kompas T. and Schnieder, V., 'The Bioeconomics of Marine Reserves: A Selected Review with Policy Implications', *Journal of Bioeconomics*, 7 (2005): 161-78.

Grafton, R.Q. and Nelson, H.W., *The Effects of Buy-back Programs in the British Columbia Salmon Fishery* (Paper presented at the International Workshop on Fishing Vessel and License Buybacks Programs, Institute of the Americas, University of California, San Diego, March 22-24 2004).

Grafton, R.Q. and Silva-Echenique, J., 'How to Manage Nature? Strategies, Predator-Prey Models, and Chaos', *Marine Resource Economics*, 12 (1997): 127-43.

Grafton, R.Q., Squires, D. and Kirkley J.E., 'Private Property Rights and the Crises in World Fisheries: Turning the Tide?', *Contemporary Economic Policy*, 14 (1996): 90-9.

Gulland, J.A., *Fish Stock Assessment: A Manual of Basic Methods* (Chichester: John Wiley & Sons, 1983).

Hannesson, R., *Fisheries Mismanagement: The Case of the North Atlantic Cod* (Osney Mead, UK: Fishing News Books, 1996).

Hardin, G., 'The Tragedy of the Commons', *Science*, 162 (1968): 1243-48.

Hilborn, R., Branch, T.A., Ernst, B., Magnusson, A., Minte-Vera, C.V., Scheurell,

M.D. and Valero, J.L., 'State of the World's Fisheries', *Annual Review of Environment and Resources*, 28 (2003): 15.1-15.40.

Hilborn, R., Orensanz, J.M., and Parma, A.M., 'Institutions, Incentives and the Future of Fisheries', *Philosophical Transactions of the Royal Society*, 360 (2005): 47-57.

Hilborn, R. and Walters, C.J., *Quantitative Fisheries Stock Assessment: Choice, Dynamics and Uncertainty* (New York: Chapman and Hall, 1992).

Holland, D., Gudmundson, E. and Gates, J., 'Do Fishing Vessel Buyback Programs Work: A Survey of the Evidence', *Marine Policy*, 23 (1999): 47-69.

Kompas, T. and Che, T.N., *Management Options under Uncertainty: A Bioeconomic Model of the Australian Northern Prawn Fishery* (Canberra: ABARE, 2004).

Larkin, P.A., 'Fisheries Management - An Essay for Ecologists', Annual Review of Ecology and Systematics, 9 (1978): 57-73.

Morey, E.R., 'A Generalized Harvest Function for Fishing: Allocating Effort among Common property Cod Stocks (A Generalized Harvest Function)', *Journal of Environmental Economics and Management*, 13 (1986): 30-49.

Pauly, D., Christensen, V., Dalsgaard, J., Froese, R., and Torres, F.Jr., 'Fishing Down Marine Food Webs', *Science*, 279 (1998): 860-3.

Pauly, D., Christensen, V., Guénette, S., Pitcher, T.J., Sumaila, U.R., Walters, C.J., Watson, R., and Zeller, D., 'Towards Sustainability in World Fisheries', *Nature*, 418 (2002): 689-95.

Pella, J.J. and Tomlinson, P.K., 'A Generalized Stock Production Model', *Bulletin of the Inter-American Tropical Tuna Commission*, 13 (1969): 419-96.

Pikitch, E.K., Santora, C., Babcock, E.A., Bakun, A., Bonfil, R., Conover, D.O., Dayton, P., Doukakis, P., Fluharty, D., Herman, B., Houde, E.D., Link, J., Livingston, P.A., Mantel, M., McAllister, M.K., Pope, J., and Sainsbury, K.J., 'Ecosystem-Based Fishery Management', *Science*, 305 (2004): 346-7.

Ricker, W.E., 'Stock and Recruitment', *Journal of the Fisheries Research Board of Canada*, 11 (1954): 559-623.

Schiermeier, Q., 'How many more fish in the sea?', *Nature*, 419 (2002): 662.

Townsend, R.E., 'Entry Restrictions in the Fishery: A Survey of the Evidence', *Land Economics*, 66 (1990): 359-78.

Wang, Y-G and Die, D., 'Stock-recruitment Relationship of the Tiger Prawns (*Penaeus Esculentus and Penaeus Semisulcatus*) in the Australian Northern Prawn Fishery', *Marine and Freshwater Research*, 47 (1996): 87-95.

Wilen, J., 'Fisherman Behaviour and the Design of Efficient Fisheries Regulation Programmes', *Journal of the Fisheries Research Board of Canada*, 36 (1979): 855-8.

Wilson, J.A., Acheson, J.M., Metcalfe, M. and Kleban, P., 'Chaos, Complexity and Community Management of Fisheries', *Marine Policy*, 18 (1994): 291-305.

Chapter 2

Data for Economic Analysis of Commercial Fisheries

You cannot do good analysis if the data are not good.
Ray Hilborn and Marc Mangel (1997, p. 10) in The Ecological Detective: Confronting
Models with Data.

2.1 Introduction

This chapter addresses collection, survey design, and the use of data for economic analysis of commercial fisheries. In particular, we consider the issues of data collection in both developed and developing countries, and for both large-scale commercial fisheries and small-scale or artisan fisheries. Such data provide the basis for good decision-making and better fisheries management. Our focus is on the type of data to be collected, how it is collected and its implications for economic analysis that are examined in details in chapters 3, 4 and 5.

2.2 Unique Challenges of Economic Data Collection for Fisheries

Economic analysis of fisheries is used to support the conservation and management of fisheries industries, habitat, and protected species such as sea turtles, dolphins, whales, and sea birds. Fisheries economists study the economic activities of harvesting, processing, and the effects on natural resource stocks and environmental assets. This forces them to confront data issues that extend beyond those typically faced by economists who analyze more conventional industries and household consumption. For example, economic analyses at the individual vessel level of fisheries that harvest highly migratory species, such as swordfish and tuna, not only requires conventional economic data on inputs, outputs, and their prices, but additional data on the environment, such as sea surface temperature and on the abundance of the stocks of fish.

The multi-product nature of many fishing industries also poses other challenges for data collection. In fisheries the general focus on landings necessitates a multi-product analysis to capture the nature of joint production and the economic and technological relationships between individual species and individual inputs. In addition, in some fisheries there may be bycatch of species that are of little commercial value, or inadvertent production of public or common 'bads', such as mortal harm to protected species or habitat destruction. In most instances, these undesirable outputs are produced at sea, and hence are unobserved. A comprehensive economic analysis in these situations requires collecting data on not only the desirable outputs, such as target species, but also the undesirable by-products. Because bycatches are typically discarded at sea, special efforts are required through on-board observers to collect the relevant data.

In fisheries, biologists almost always establish the fundamental data collection systems and the production process is at sea away from easy observation. Consequently, when data collection systems are established, they typically focus on monitoring the status of the resource stocks by collecting landings data on the different species caught and perhaps the length, weight, ages, and species composition beyond commercial market categories. This very much leaves economic data collection as an 'after thought' and analysts are frequently obliged to make the best use of the existing data, or of periodic ad hoc data collection efforts, rather than on-going systematic collection of information better suited to the needs of economic analysis.

2.3 Priorities in Data Collection

Given limited budgets and paucity of existing economic data in fisheries, analysts must make the fullest possible use of ongoing data collection efforts, usually biologically based. Systematic economic data collection is also least costly if it exploits existing data collection infrastructure. To the extent that port sampling and landings data collection or observer programs can provide the most important trip costs and input quantities, costs are minimized with the joint production of both biological and economic data. This requires, however, that only the most essential economic data be collected rather than the full, detailed suite that economists might wish to collect.

Greater cooperation in data collection with fishers and processors is also required to make the best use of scarce resources. For example, fishers and processors subject to conservation and management policies have an inherent interest in making available the best possible scientific information, and can often be persuaded to readily cooperate with economists in collecting such information. Attention should also be given to getting the most out of any data that is collected and making effective use of the available information.

2.4 Types of Economic Data

Cross-Sectional Data

Economic data are generally of two types, cross-sectional and time-series. In cross-sectional data there are observations on individual units at a point in time. For example, a cross-sectional data set used to estimate a fishery production function might include annual observations for fishing vessels on catch, vessel size (such as gross registered tons), crew size (including captain), and fishing time (such as days fished, capturing variable input usage as a proxy variable). These data could come from a landings data set, in which all landings of vessels in a fishery are recorded in a given year, or they could come from sample surveys. Another type of cross-sectional data is more aggregated, such as a cross-section of states, regions, or countries. In some instances, the data on all units do not correspond to precisely the same time period. For example, different fishing vessels may be surveyed during different weeks within a year. Data collected from income tax returns or income statements may confront different beginning and ending dates for tax years. In artisan fisheries, where surveys rely upon recall of recent fishing trips, fishers in different areas may be surveyed in different weeks or even seasons during a year. In a pure cross-sectional analysis, minor timing differences in collecting the data would be ignored, but more substantive differences, such as data collected during different parts of the year, should not be overlooked.

An important feature of cross-sectional data is the assumption that the data have been collected from a random sample of the underlying population (Wooldridge 2002). For example, suppose we are interested in using landing records to obtain information on costs and earnings and other characteristics of vessels in a fishery. Landings records of all vessels in the fishery constitute the population, or the set of all potential observations. A sample drawn from the population where each boat is equally likely to be included constitutes a random sample. If the most accessible vessels were more likely to be surveyed, the data would not constitute a random sample of the population.

Violations of random sampling can occur (Deaton 1997, Wooldridge 2003) due to factors that favor the inclusion of some individuals in the population over others. For instance, some group in the population may be less likely to provide sample information such that a sample of cost and earnings of fishers is not a random sample of all fishing vessels. Violation of random sampling occurs when samples are drawn from units that are large relative to the population without adjusting the sampling procedure for the relative size of the sampled units. In such cases, the population may not be large enough to reasonably assume that the observations are independent draws from the population. For example, more profitable fishing vessels may be clustered around the most abundant fishing grounds with higher-valued species. In this case, it is unlikely that vessel activities in ports near these grounds are independently representative of the overall population.

Time-Series Data

Time-series data contain observations over a period of time on a variable or several variables. These data typically are highly aggregated and often are compiled over all firms, markets, or consumers in a given time period rather than at the level of the individual economic agent (firm, vessel, consumer, household). Two widely used sets of time-series data in fisheries economics are catch and effort data, and price data. Time series of catch and effort data are typically used to estimate bioeconomic models (see chapter 1).

Intertemporal dependence is an important feature of time-series data (Wooldridge 2003), because past events can influence future events and lags in behavior are prevalent in fishing industries. In addition, some time-series data may also include some trend. If this is the case, then a time-series relationship between catch and effort might indicate that a spurious relationship exists between the two variables. The chronological ordering of observations in a time series might also convey potentially important information to analysts. By contrast to cross-sectional data, economic observations in time-series data are rarely, if ever, assumed to be independent of one another across time (Wooldridge 2003) because most economic time series are related to their recent histories, such as the ex-vessel price of a species from one day to the next.

The frequency at which data are collected is another important feature of time-series data. The most common frequencies for economic data in most industries are daily, weekly, monthly, quarterly, and annually. In fisheries, however, landings and prices are often collected on the basis of a fishing trip or days fishing or at sea. Because of the close relationship between fishing activity and seasons of the year, strong seasonal patterns are often apparent in a time series of economic variables in a fishery.

Pooled Cross-Sections

A third basic type of economic data is a series of cross-sectional data samples chosen at different points in time, which is sometimes called pseudo-panel data or cohort data. Such data sets have both cross-sectional and time-series aspects. An example is the series of annual cross-sectional surveys of the British Columbia halibut longline vessel cost-and-earnings conducted by the Canadian Department of Fisheries and Oceans, which were used by Grafton et al. (2000). Each cross-sectional data set was collected on the basis of a random sample of the fleet in the year that the sample was collected, but for each year the sample selected was independent of the previous years.

Panel or Longitudinal Data

A fourth basic type of economic data are panel data that are repeat observations made on each cross-sectional unit in a data set over different time periods. The data units can be individual economic agents, such as consumers, households, fishers, processors, or vessels, or aggregate cross-sectional units at the regional or national level. Because panel data require replicated observations of the same units over time, they are often more difficult to obtain than pooled cross-sections.

Observing the same units over time offers several advantages over cross-sectional data or pooled cross-sectional data. Multiple observations on the same units allow the analyst to control certain unobserved characteristics of individuals, firms, households, and so on. Multiple observations on individual units through time can facilitate causal inference in situations where inferring causality would be difficult with cross-sectional data from a single period. For example, Squires and Kirkley (1999) used a panel of US Pacific coast groundfish trawlers to quantify the unobserved skipper skills of these vessels, and by controlling for this source of unobserved heterogeneity in management skill, were able to more accurately evaluate substitution possibilities among different vessel inputs, such as capital, labor, and fuel. Panel data also enable the researcher to study the importance of lags in behavior or the results of decision-making.

Panel data can be balanced or unbalanced. Balanced panel data occur when data are available for each cross-sectional unit and time period represented in the data. By contrast, unbalanced panel data have missing values for some cross-sectional units in some time periods that are included. There are also a number of variations on how to establish panel data, such as rotating panel data, where the units of interest are rotated out of the sample based on pre-specified rules. A rotating panel balances the advantage of observing the same individuals across multiple periods against the desire to keep the sample representative of the population over time by allowing sampling new entrants into the panel.

2.5 Experimental and Non-Experimental Nature of Data

In controlled experiments the variables hypothesized to influence a dependent variable can be set in advance and the response by the dependent variable observed. Unfortunately experiments are seldom possible in economics, but quasi or natural experiments can shed important light on policy events. As a result fisheries economics uses non-experimental survey or landings data to look at the differences in behavior between economic agents, often by individual vessels, and explores the relationship between degree of 'exposure' to the treatment and variation in the outcomes of interest.

A related data issue is whether the data on the policy event, or 'treatment' to be analyzed, come from a representative sample of the population. Many investigators treat the data as though it measured replications of a controlled experiment subject to the researcher's exogenous choice of control levels. In reality, the data typically do not come from controlled experiments in which the effects of a treatment can be convincingly determined or confounding influences ruled out (Deaton 1997). Explanatory variables are, thus, prone to be endogenous in the sense that they may be influenced by the variable they are purported to explain. This is likely to give rise to estimation problems in statistical models because it can result in correlation between explanatory variables and the random error term. Although this can be mitigated using certain methods, it can lead to biased estimators where the expected value does not equal its true value, and inconsistent estimators such that the bias does not tend to zero as the sample size increases.

Ideally, to evaluate the effects of a policy or 'treatment,' an experiment would be conducted in which some randomly chosen group is given the 'treatment' and the results compared with a randomly selected group of controls from whom the treatment is withheld. Randomization guarantees that there are no differences between the two groups aside from chance variation, so that if there is a significant difference in the average outcome, it is most likely due to the effect of the 'treatment'.

In the absence of natural experiments, standard survey techniques are often used as the basis for evaluating policy effects, especially to obtain cost data in developed countries or comprehensive economic data in developing countries. In developed and some developing country fisheries, landings data collected by government agencies are a major source of data. Both survey and landings data – even when the former are collected by a random sample and the latter are population data (of all landings) – are non-experimental in nature.

Many of the empirical microeconomic analyses in fisheries use econometric and statistical methodology that reflects the non-experimental nature of the data. This enables the analyst to relate the degree of exposure to a treatment or policy and variation in the outcomes of interest without the need to conduct controlled experiments. Statistical assumptions are made to bridge the difference between theory and data, thereby justifying estimation of model parameters and the subsequent interpretation of the data. However, if the assumptions cannot be verified, neither can the applicability of the theory used to draw conclusions.

Non-experimental Data and Endogeneity

Fisheries economists frequently use data to estimate a statistical relationship between a set of explanatory variables and a dependent variable of economic interest. This is called econometrics because it combines an economic model with a statistical model to provide insights about an economic relationship. For example, a production function that maps inputs into an output might be estimated using individual vessel data on inputs (vessel size, number of crew, etc.) as the explanatory variables (defined as a vector x), and vessel landings, defined by y, as the dependent variable. Typically, at least one of the variables comprising the explanatory variables is a 'treatment' variable while others are 'control' variables. The control variables allow for differences in outcomes not caused by the treatment. This helps the analyst to isolate the treatment effect and allow the control variables to play a role similar to that of a control group in a scientific experiment. For example, when estimating a production function for a fishery a control variable would be the size of the fish stock.

The error (or residual or disturbance) term in an econometric model, u, captures omitted controls and measurement error on the dependent variable. The error term is assumed to satisfy the requirement that its expectation is conditional on the explanatory variables, denoted by x, is 0, i.e., $E(u \mid x) = 0$. This assumption states that the average value of the random error, defined by u, does not depend on the value of x. That is, for any given value of x, the average of the 'unobservables' is the same and, therefore, must equal the average value of u in the population.

The most common problem with econometric analysis is the failure or implausibility of the assumption that $E(u \mid x) = 0$. If the explanatory variables are correlated with the error term, they are said to be endogenous. Several problems can lead to violation of this assumption, including omission of a relevant variable that is correlated with any of the included variables in the vector of explanatory variables x, error in measurement of the explanatory variables, and simultaneous causality between x and y. When this occurs the econometric model no longer coincides with the structure of interest, and estimation yields biased estimators of the model parameters (Deaton 1997, Wooldridge 2002, 2003).

In the omitted variable case, one or more of the appropriate explanatory variables is omitted from the regression equation. The effect of an explanatory variable which is correlated with both the dependent variable y and some of the explanatory variables in the vector x shows up in the error term, thereby violating the assumption that $E(u \mid x) = 0$.

Simultaneity between the x and y variables arise when one or more of the explanatory variables comprising the vector x are not pre-determined, and instead are simultaneously determined by factors that jointly influence y. The error term u is then correlated with one or more of the variables comprising x and standard statistical estimates will be biased and inconsistent.

Besides omitted explanatory variables and simultaneous determination of y and x, a third reason for $E(u \mid x) \neq 0$ and biased and inconsistent econometric estimates of model parameters is measurement error, also called errors-in-variables (Wooldrige 2002). Thus, when collecting economic data, it is important to collect data on all of the important variables that are likely to be important and to keep measurement error to a minimum.

Natural Experiments

Natural (or quasi natural) experiments allow for distinct groups of individuals to receive different treatments in a non-experimental setting in a similar manner to the randomized assignment of treatments to subjects of a controlled experiment. Natural experiments also permit researchers to measure the impact of a policy change on sets of individuals who are treated differently across time periods, or spatial units, without the need to conduct a survey or to carry out a designed experiment. Although the treatments are not randomly assigned to subjects by design, a natural experiment enables the researcher to proceed as if this were the case (Deaton 1997, Wooldridge 2002).

Grafton et al. (2000) exploited the concept of a natural experiment to analyze time series of cross-sections from before and after the introduction of individual transferable quotas in the British Columbia halibut fleet. In this case, data were collected as individual cross-sections of data from three different years, but were combined to evaluate the effects of ITQs on economic efficiency. The 'treatment' variable was the introduction of transferable property rights for catch into what was previously a restricted or limited open-access longline fleet. Because the data were pseudo-panel, i.e., a time series of cross-sectional data, Grafton et al. (2000) analyzed cohorts of vessel size classes rather than individual vessels.

Natural experiments are not genuine, randomized experiments, and Deaton (1997, p. 114) observes that a case-by-case argument has to be made that the natural experiment is effectively randomized. If differences in treatment are not random, but instead are linked to some characteristic of the fleet or area this will invalidate the natural experiment interpretation. In addition, the ideal natural experiment analysis would include observations on both a control group and an experimental (treatment) group (Wooldridge 2002) from before and after exposure to the treatment. In the Grafton et al. (2000) study, all halibut longline vessels were evaluated as part of the treatment group, because there were no halibut longline vessels that did not receive transferable property rights to form the control group in this case. An ideal natural experiment in this case would have included data from before and after the introduction of ITQs for a treatment group of vessels receiving the transferable property right, and an incidentally random control group of vessels that did not receive the rights.

Panel (Longitudinal) Data

Panel (longitudinal) data are comprised of repeated observations on each individual, allowing comparison of the same individual under different conditions. Panel data also allow econometric estimation that controls for individual heterogeneity, improve efficiency by using data with more variability, add more degrees of freedom, and reduce collinearity or correlation among the explanatory variables. Their use facilitates estimation and testing of more complicated behavioral models, and examination of adjustment dynamics (Hsiao 1986, Baltagi 1995, Arellano and Honoré 2001). Panel data also allow for identification and measurement of effects that are undetectable in pure cross-sections or pure time-series data, such as turnover in vessel composition of a fleet. For example, Squires and Kirkley (1999) exploited panel data to evaluate skipper skill in the US Pacific Coast trawl fishery to depict otherwise unobserved managerial or skipper skill by inter-vessel differences.

Despite the potential benefits of panel data, selectivity problems can arise when individuals self select out of the sample (Baltagi 1995, Wooldridge 2003). Non-response occurs at the initial wave of the panel, when individuals refuse to participate, are not at port, are an untraced sample unit, or for other reasons. Non-response is less of a problem in cross-sectional studies than in panels because subsequent waves of the panel are still subject to non-response, but the pattern of non-response in the panel may vary over time.

Attrition is also an issue, as vessels or processors leave the fishery over time and the remaining sample is less representative of the original population. Respondents may exit the fishery due to non-profitability, a move to another fishery or because the cost of responding is too high. Rotating panels offer a response to attrition by holding the number of economic agents in the survey constant by replacing those leaving the sample each period by an equal number of freshly surveyed individuals. New entrants that are not captured in the panel pose a similar problem. Too few time periods of observation on each individual in the panel can also pose a difficulty in terms of data analysis.

Pooled Cross-Sections over Time: Pseudo-Panel or Cohort Data

Long-running panels are rare, but independent cross-sectional surveys over time are common. For example, the Department of Fisheries and Oceans in Canada conducts cost-and-earnings surveys on a periodic basis, but draws a new random sample each time the survey is administered, rather than following the same vessels over time. Combining these independent cross-sections of data obtained over time into one data set creates pseudo-panel or cohort data. The idea is that during each time period a new random sample is taken from the relevant population.

Pseudo-panel data can be created out of cross-sections of individuals over time by aggregating over the individuals for desired cohorts or categories of analysis, such as vessel size classes, gear types, or locations. In this case, econometric analysis is properly conducted on the cohorts constructed by aggregation, such as vessel size classes, rather than individual observations (Deaton 1985). For example, in an evaluation of economic efficiency from the introduction of individual transferable quotas into the British Columbia halibut longline fishery, Grafton et al. (2000) combined three independent cross-sectional cost-and-earnings surveys.

A problem can arise with the assumption that the cohort population is constant, which is required if the surveys are to be treated as successive random samples from the same underlying population. In addition, the cohort data do not allow analysis of the dynamics within the cohorts. Instead, each survey provides information about the distribution of the characteristic in the cohort in each period, but two adjacent surveys indicate nothing about the joint distribution of the characteristic in the two periods (Deaton 1997). Other than dynamics, the cohort data allow much of the same analysis that could be conducted using actual panel data. Most importantly, cohort-level data implicitly control for sources of unobserved heterogeneity among individual economic agents, such as fishing skill or individual consumer tastes and preferences, and thereby allow consistent econometric estimates of parameters.

Cohort (pseudo-panel) data have several advantages over panel data (Deaton 1997). Cohort data are constructed from fresh samples every year, so that attrition is not an issue. However, changes in sampling design over time can affect the 'representativeness' of the cohort that is analyzed. Analysis of cohorts is also less likely to be affected by measurement error. The cohort that is tracked over time, such as a vessel size class, is usually an average, median (useful with outliers), percentile, or some other similar summary statistic. These summary statistics nearly always have smaller measurement error, since the averaging reduces the effects of measurement error and increases the signal-to-noise ratio.

2.6 Aggregate and Microeconomic Data

Data are often available only at a relatively high level of aggregation. Fishery data, in particular, can range from the highly aggregated, such as those published by the Food and Agriculture Organization of the United Nations (FAO) and many national governments, to highly disaggregated data at the level of the set of the gear used on an individual vessel. Data can also be aggregated at the national, regional, or port levels and over gear types and vessel size classes. Data can also be aggregated over time periods, ranging from the individual deployment of gear, such as set of the net, to an annual basis or an intermediate state such as daily, weekly, monthly, or quarterly, or some other natural season. Data can also be aggregated over outputs, ranging from total catch, to catch by species or even by market categories for an individual species.

Individual variation in the data are naturally lost the higher is the degree of aggregation across firms, households, or time periods. Aggregation to some degree, however, can be useful as a means to 'smooth' the inherently stochastic nature of fisheries data, and even to cancel out measurement errors. For example, econometric estimation of models of a vessel's harvest at the set level is heavily influenced by stochastic events beyond the control of the captain and crew, such as randomness in the location of fish and the presence of micro-climates or storm fronts. The shorter is the time period and more tightly focused the geographic area of aggregation, the more that environmental variables and stochastic events will influence the econometric results. Multiproduct production models with a highly disaggregated output vector (catches of individual species), and specified at the level of the individual set, may also fail to display desirable properties required by analysts, but more highly aggregated outputs and time periods may have these desirable properties. Mathematical programming models, which are generally less adept than econometric models at handling multiple time periods (such as panel or pseudo panel data), frequently aggregate data over multiple years in order to smooth the individual and idiosyncratic effects found in a given year. Any particular year may not be considered representative of the fishery by fishers or managers, but annual data averaged over two or three years may be considered more representative of the fishery over time. Such aggregation also smoothes out the annual effects of weather, environmental influences, other stochastic events, and measurement error.

Highly aggregated data that are combined across all vessels in a fishery and then used by analysts inherently assume an aggregate production technology or consumption function. Aggregate consumption is treated as if it had been generated by the decision processes of a single, 'representative agent' (Deaton 1992). Similarly, aggregate production is treated as if it had been generated by the decision processes of a single, 'representative' firm, or in the case of fisheries and bioeconomic modeling, a 'representative' vessel.

Specification of aggregated or disaggregated production or consumption models (including bioeconomic models) raises the issue of consistent aggregation across outputs, inputs, and consumer goods. Data that are aggregated across species of fish, different types of inputs, or different consumption categories implicitly assume certain characteristics of the underlying data. When data are collected, it is often advisable to begin with disaggregated data and to aggregate as required by categories that are meaningful, using the appropriate aggregation procedures. For example, in an analysis of a groundfish fishery, various species of flounders and soles could be aggregated into flat fish, cod could be left as cod, haddock could be left as haddock, and other fish might be aggregated into a composite group, others. The various species grouped into a single output, such as the different species of flounders and soles aggregated into flatfish, should be aggregated by an index number procedure using some form of weight such as revenue shares (see chapter 5).

Collection of data representing fishing effort is a particularly difficult aggregation issue in fisheries. One interpretation of fishing effort is that it represents a composite input comprising all inputs (Hannesson 1983, Squires 1984). Fishing effort is theoretically formed as part of a separable two-stage production process, whereby in the first stage of production fishers optimally and efficiently combine factors of production, such as capital, labor, and energy to form the composite input index, effort. A second interpretation is that fishing effort is the product of a non-separable, two-stage production process (Pollock and Wales 1987, Squires and Kirkley 1991). A third interpretation is that fishing effort is a composite input of variable inputs.

Frequently, and often out of necessity, the only data representing fishing effort is a measure of fishing time, such as total days at sea, perhaps multiplied by some measure of vessel size or number of vessels to provide a flow of services. The use of fishing time to represent fishing effort as all inputs implicitly makes assumptions about the relationship between the underlying components of fishing effort. Use of fishing time to represent the unobserved variable fishing effort also represents the use of a proxy variable. The proxy variable by definition contains measurement errors, and thus biases econometric estimates. Forgoing the proxy and simply omitting the unobservable variable also creates bias in statistical estimation (omitted variable bias) if the omitted variable is correlated with the included explanatory variables.

Survey data may be presented in the form of a table summarizing the values for individual observations. Group sums or group means may be used instead of individual observations for econometric models. This aggregation of data raises the issue of 'grouped' data. In principle, aggregating data and analyzing group means or sums lowers efficiency because of the loss in information, although in practice, the loss could be less than expected since aggregation to some degree cancels out errors in measurement or misspecifications of micro-relationships. Heteroscedasticity, or correlation between the error term and one or more explanatory variables, arises if each group does not contain the same number of observations even if the original errors were not correlated with the explanatory variables (Johnston 1972). Care must also be taken in determining how the groups are chosen because different results can be obtained with different grouping rules (Maddala 1979).

Aggregating Individual Outputs, Inputs, or Consumption Goods by Index Numbers

When aggregating quantity data on individual species or other outputs, inputs, and consumptions goods, simply adding the quantities together for the category under consideration — called linear aggregation — can impose a bias on the aggregated good for subsequent economic analysis. Instead, various methods or formulae exist by which to aggregate the goods of concern. These methods or aggregator functions are called economic index numbers and are discussed in chapter 5.

2.7 Administrative Data

Administrative data are regularly collected by organizations for regulatory reasons rather than analytical purposes. Administrative data can provide a very close approximation to the entire population, but in other cases, they comprise samples that are not randomly drawn from the population and are hence not representative. There are at least five types of administrative data that are used in commercial fisheries.

Landings data

Data on commercial fisheries landings are routinely collected in many developed countries when vessels arrive in port and land their catch. On the west coast of the continental United States, for example, when vessels sell their catch to fish buyers (processors or at auctions), a 'pink slip' is completed with such information as ex-vessel sales revenue and quantity of landings by species, date of trip departure and arrival, and port. The quantity and revenue data can be assumed accurate, with minimal measurement error, since the data constitute the official records of commercial transactions between fishing vessels and processors.

In many instances, the landings are described by market categories rather than by species, and as such, individual species are often lumped together into a general category. For example, on the west coast of the United States, various species of rockfish are simply classified as 'rockfish' rather than by their individual species, of which there are many. In some instances, government biologists systematically sample the landings data for further information on species composition and to collect length and weight information. These data are collected by states and then frequently aggregated across states by the US federal government. These landings data can be viewed as population data (where the population in question is the overall US commercial catch) because (almost) all of the commercial vessels' landings data are collected. Landings data, as population data, are particularly valuable when designing sample surveys, including sample size and stratification, where the sample surveys may be intended to collect cost data which are not gathered at the time of landing.

Landings data, however, do not necessarily completely reflect catches as some harvest may be discarded at sea. The seriousness of this issue varies by fishery, time period, and regulatory regime. One hundred per cent observer coverage is the only way to accurately measure at-sea discards and ensure that all catch is recorded. For example, the British Columbia multi-species trawl fishery, which is regulated by individual transferable quotas, imposes one hundred per cent observer coverage to insure that all catches at sea are accurately recorded (Grafton et al. 2004).

Census Data

Censuses are data taken by central governments on a regular basis of the general population. These data can be useful in analyses of consumer behavior, in economic cost-benefit analyses of fishery management actions, such as constructing economic opportunity costs for crewmembers, and in construction of economic impact multipliers. As population data, census data can also be useful when designing sample surveys by providing the information by which to calculate sample size.

Census data may also include data on vessel characteristics that are routinely collected by governments. For example, when applying for permits, information on selected vessel characteristics is often requested. The US Coast Guard, for instance, collects information on vessel characteristics, such as gross registered tons and registered vessel length. Similarly, the Forum Fisheries Agency and the Secretariat for the Pacific Community compile information on vessel characteristics for tuna purse seine vessels from different flag states fishing in the Western and Central Pacific Ocean.

Observer Data

Some fisheries are observed for scientific purposes or to collect information on bycatch, discards, takes of protected and endangered species, and monitoring in general. For example, the California drift gillnet fishery interacts with, and incidentally catches, marine mammals and protected species such as sea turtles and sea birds. The fishery is covered by approximately 20 per cent observer coverage. Some vessels in the international tropical tuna purse seine fishery in the Eastern Tropical Pacific find and subsequently pursue mature yellowfin tunas by first sighting dolphins and then circling schools of dolphins because the large yellowfin tuna tend to aggregate under the dolphin schools (Joseph 1994). The International Dolphin Observer Program was formed to address the dolphin bycatch and contains one hundred per cent observer coverage. Observers typically collect routine biological data, but can also collect cost data. In some instances, observers also record environmental conditions, such as sea surface temperature, ocean conditions, or wind force. Observer data on catches and bycatches are often in the form of numbers of fish, whereas economic analysis is usually in terms of volume of fish. In these instances, numbers of fish must be converted to volume using other observer information on mean weights per fish.

Observer data collected by programs with one hundred per cent observer coverage can be viewed as population data. In this case, there is not an issue in terms of the representativeness of the data, nor are there concerns over statistical inferences about the population from a sample and the accuracy of the sample. When observer data are collected with less than one hundred per cent observer coverage, however, issues arise as to whether fisher behavior is the same under observation as when there is no observer coverage. Even when observers are purely for scientific reasons, observers perform a monitoring function and even indirectly may help enforce conservation and management measures. Vessels receiving observer coverage may also not be representative of the entire population of vessels. For example, only vessels of sufficient size can carry observers.

Measures computed from observer data, such as total catch, are typically simply extrapolated to provide population estimates. For example, in the California drift gillnet fishery with 20 per cent observer coverage, total catches are multiplied by five to provide the population estimate. This practice, while perhaps inevitable, raises questions when extrapolating very rare events calculated from the observer data, such as sea turtle take and mortality rates, to the population of all vessels and trips.

Vessel Logbooks

In many fisheries vessel captains maintain detailed logbooks on each fishing trip of location fished, catches (numbers of fish or volume) by species, bycatches, environmental conditions such as sea surface temperature, wind force, and weather, various measures of fishing time, such as days at sea and days fished, identity of the captain, number of crewmembers, port of departure and landing, and other such related data. However, logbook data do not contain price information and values entered into logbooks, such as catches, are frequently estimates at the time the gear is brought on board and unloaded or the fish unhooked, and as such may be subject to measurement error.

Logbook information is inherently at the level of the individual trip. Often, logbook information is the only source of data for a fishery, unless landings data are systematically collected by a government agency. The number of vessels entering logbook data varies. Hence, whether or not the population of vessels completes the logbooks is determined on a case-by-case basis. When most, but not all, vessels complete the logbooks, how representative this non-random sample is of the population must also be determined. Another issue with logbook data is gathering the data and subsequently entering it electronically. Often, because of the size of this task, a considerable time lag occurs between the collection of logbooks at the end of the fishing season or year and the availability of data in electronic form.

Auction Markets

Auction markets are sometimes used to assemble landings and establish sales of fish. The fish auction markets in Sydney, New York (Fulton), and Tokyo (Tsukiji) are well known. Other auctions are found in Marseille, Seattle, Honolulu, New Bedford, and elsewhere. Ex-vessel prices formed in these markets for various species are frequently collected by public agencies, and have served as the basis for many econometric analyses of spatial price linkages and market integration. Not all fish caught in waters supplying fish to auction markets necessarily enter the auction market, so that in these instances, care must be taken if the only source of price data are auction market data.

2.8 Survey Data

Economic data on fisheries can be collected by surveys of vessels, processors, brokers and distributors, or even households. Surveys in developed countries generally tend to focus on collecting cost and input data because landings data from entire fleets are often available to provide catch and revenue (and thereby price) information. Landings data also frequently provide some measure of fishing time, ranging from number of trips to total days at sea to days fished. Vessel registers also record various measures of vessel size, such as some measure of length, gross tonnage, or carrying capacity, and sometimes main engine horsepower or kilowatts. Vessel registers typically are less comprehensive on other measures of inputs, such as crew size, fuel consumption, and gear and equipment.

Surveys in developing countries, where collection of landings data is typically less extensive, often aim at collecting data on both the input and output sides, that is, on physical measures of inputs, costs, landings, and revenues. Surveys of artisan fishers in developing countries may also collect information on other production processes emanating from the fishing household, such as agriculture, gathering, informal marketing, or formal marketing often performed by spouses of the artisan fishers, and so forth. In addition, a certain proportion of the fish harvested may not enter the market, but instead are consumed by the household or informally exchanged within the village or hamlet for other products. Surveys of artisan fishers may also collect information about economic relationships with fish purchasers, who may extend credit to the fishers in return for landings of fish at ex-vessel prices that may fall short of market prices. In these instances, ex-vessel prices reflect payments for credit and assumption of risk by the fish brokers. Surveys in developed and developing countries alike are often not part of an ongoing data collection process, but are instead administered on an occasional basis.

There are many types of survey data: cross-sectional, supplemental cross-sectional data, panel, rotating panel, time series, time series of cross-sections, and opportunistically collected data. This section discusses each of these types of data and the issues each of them raises for analysis and survey design.

Sample Size

When collecting data, particularly from surveys, or assembling data sets from landings data, a critical issue is that of sample size. Sample size can affect the subsequent econometric estimation and the accuracy and precision of sample statistics (such as means). The sampling distribution of most estimators changes as the sample size changes. For example, one of the most important summary statistics, the mean of a simple random sample, has a sampling distribution centered over the population mean, but a variance that becomes smaller as the sample size increases. The precision of sample statistics increases less than proportionally with the sample size (usually in proportion to its square root), so that sampling fractions are typically smaller in large populations.

When designing a sample survey, choice of sample size is also related to the desired level of accuracy (allowable size of error from the population parameter) and the level of precision (sampling variation) of estimated sample statistics. For example, given the characteristics of the population to be surveyed, a sample of a certain size will give a sample mean that has 10 per cent error from the population mean at a level of precision of 5 per cent (i.e., a 5 per cent level of significance). Thus, given a population mean of 100, a sample of some size (depending on the population surveyed) would have a sample mean that is within 90 per cent to 110 per cent of the population mean 95 per cent of the time. Choice of sample size also depends on the type of sample, whether the sample is selected using simple random sampling, systematic sampling (which involves listing the population and then picking every tenth or every hundredth, etc. observation), stratified sampling (population is divided into groups called strata and random samples are chosen from the strata), cluster sampling or multistage sampling (population is divided into groups which are then subdivided and random sampling of the subdivisions is made and then after a random sample of observations is made from the picked subdivisions).

A relatively large sample size also helps justify an appeal to the Central Limit Theorem, which helps ensure sample statistics and parameter estimates that depend on sample means are approximately normally distributed in samples of reasonable size. In other words, large samples provide some assurance that the sample mean and sample standard deviation, for example, are approximately normally distributed about the population mean and population standard deviation. Normally distributed estimators for econometric models also permit the application of most statistical analysis, such as F-tests and t-tests. A larger data set can also ensure that asymptotic properties of estimators are more likely to hold. One of the most important of these large-sample properties is consistency, where the probability the estimator will assume a value close to the parameter increases with sample size.

Sample size can also have an effect on the results from hypothesis testing (Leamer 1978, pp. 100-20, Deaton 1995, p. 130). Classical statistical procedures set the critical value for a hypothesis test in such a way that the probability of rejecting the null when correct, the probability of Type I error, is fixed at some pre-assigned level, such as 5 per cent. In practice, the null hypothesis is ostensibly more frequently rejected in large samples than in small. Increasing the sample size reduces the variance of the parameter of an explanatory variable and eventually makes this influence statistically significant.

Pitfalls in Surveys

(1) Sample Bias
The potential benefits of surveys are large, but there are also potential pitfalls. One major problem is sample bias. The sample may not provide an unbiased representation of the population, either due to surveying an unrepresentative cross-section, or due to non-response among those surveyed. Consequently, the sample mean may differ from the population mean. Even in larger samples, the sample mean may fail to converge upon the population mean as the sample size increases such that the estimators of the population parameters in econometric models are inconsistent.

(2) Response Bias
Response bias can arise when respondents report what they believe the questioner wants to hear. Alternatively, the respondent may attempt to influence decision-making and policy. For instance, when sample surveys are administered to groups of fishers or processors explicitly within the context of an impending fisheries management decision, such as a vessel buyback, the respondents may answer the questions in a manner they believe favors their sector of the fishery. Not only is a biased picture obtained of the fishery that may adversely affect the potential fishery management decision, but the respondents may even guess in the wrong direction by which to bias their answers!

(3) Response Accuracy
A third problem that arises with sample surveys is response accuracy. The respondent may simply be unable to answer a given question with any degree of certainty, leading to measurement error. Some questions may be hypothetical in nature, which are difficult for many people to answer. Fishers may also have little idea how they would actually react to a price increase, potential regulation (e.g., imposing a time-area closure), or some other change in regulatory or market conditions. Some questions that rely upon recall, particularly with small-scale commercial and artisan fisheries, may also lead to inaccurate responses.

Unequal selection probabilities

The more complex the sample design, the greater are the difficulties of using the data for anything other than the original purpose (Deaton 1995, 1997). Thus if survey data are to be used for a wide range of purposes then it may be best to favor simpler random samples. If not, the small gains in precision from using a stratified random sample might be obtained at the price of large compromises in the applicability of the survey to other analyses.

Estimation of means also becomes more difficult in stratified samples where the population is divided into strata or groups, because it leads to non-random samples from the population. For instance, in a simple random sample the observations in each stratum or grouping are random variables, but the fractions of the sample within each stratum will vary from sample to sample. As a result, special methods for econometric analysis of data from stratified sampling are sometimes required, because stratified sampling typically leads to non-random samples from the population (Wooldridge 2002, 2003, pp. 592-6). When different units of observation, such as households or fishing vessels, have different probabilities of being included in the survey, unweighted sample means will also be biased.

Sample Selection and Sample Selection Bias

Endogenous sample selection refers to a situation where levels of the dependent (y) variable in a regression model influence the probability that different observations will enter the sample. The consequence of this problem is that the error term in the model will be correlated with the levels of the explanatory variables, resulting in biased and inconsistent estimates of model parameters.

A classical example arises in trying to estimate the effect of years of education (the explanatory variable) on personal income (the dependent variable). If only individuals who are working (and have an income) are included in the sample a sample will under-represent individuals with low-value skill sets because they are more likely to be unemployed. If more years of education are also positively correlated with higher earnings potential, then endogenous selection of high-value workers into the sample will give rise to correlation between the error term and the explanatory variable, years of education. The standard remedy for this type of sample selectivity bias is to explicitly model the effect of the endogenous selection on the model error term (Wooldridge 2003).

Choice of Technique to Obtain a Sample

A number of questions have to be addressed in the choice of a sampling plan. First, what sampling frame or list is available? Second, how big a sample is desirable? Third, what sampling procedure is to be used? Fourth, what resources – financial, material, and human – are available for the research, and are they sufficient to meet the requirements of the sampling plan?

The choice of technique for obtaining a sample depends primarily on the nature of the problem, the cost and time factors involved, and the desired precision or reliability of the results. There is no single principle that would favor a particular sample scheme over others. The choice of technique is further complicated by the fact that in practically all surveys, the study is not limited to consideration of a single item. For example, the survey may be aimed at obtaining cost-and-earnings data, but it may additionally collect data on vessel and crew characteristics and inputs. Each of these items may call for a different sampling plan and sample size, thereby requiring a compromise on the part of the investigator as to which design to use, and how large a sample to obtain. In such instances, only those items that are essential to the study should enter into the consideration of sampling technique.

A characteristic that is present throughout the population with reasonable frequency of occurrence can be adequately investigated by means of a random sample of small size. Another item, also widespread in the population, but occurring infrequently, may be more suitably analyzed by a larger random sample. By contrast, if the characteristic is sparsely located in some parts of the survey area but concentrated in others, a stratified sample would be preferred.

Confidentiality and Response Rate Issues in Surveys

When collecting economic data from firms, such as from vessels or processors, confidentiality can be important, especially regarding cost data. Firms are understandably sensitive about competitors, tax collectors, divorce lawyers, and even regulators learning about their costs and profitability. In many instances, greater cooperation and more accurate data can be obtained by ensuring confidentiality. One such approach is to use a separate entity independent from government researchers to collect the data. Data from individual firms, plants, or vessels are then assigned an identifying number which only the collector knows, but the actual user of the data does not; this is a single-blind approach. Even greater confidentiality can be achieved by a double-blind approach, whereby the collector administers the survey instrument to individuals, but the responses are sent directly to the researcher coded with an identifier that protects the identity of the responding individuals. Higher response rates and reduced measurement error are both likely with these approaches, but at the cost of not being able to double-check the data against other, independent data sources, such as landings data.

Survey response rates, and the accuracy of answers, can also be increased by several other means. One of the most important is to directly link the reason for data collection to conservation and management requirements. Individuals are more likely to respond, and to do so accurately, when the purpose of data collection is immediate and clear. Response rates are also likely to increase when industry associations collect data. Leaders or representatives of industry associations are better known and more trusted than researchers and public officials, and are also better able to express the purposes of data collection. Survey response rates and accuracy of responses can be improved when questionnaires or other survey instruments are kept short and to the point, and when questions have been developed in conjunction with industry and with field testing. Survey response rates and accuracy can also significantly increase when respondents are monetarily compensated for providing answers.

Pilot Surveys

A final word on surveys is that the quality of the data collected is highly dependent on the survey instrument, or questions asked of the persons interviewed. While the meaning of what is being asked may seem obvious to a person who designed the questionnaire, it may mean something else to the respondent. In addition, surveys that take too long or are difficult to complete may reduce both response accuracy and the response rate. Using pilot or test surveys, prior to undertaking the actual survey, can mitigate these potential problems. A pilot survey allows the investigator to identify problems and weaknesses with the survey instrument and can be invaluable in improving the quality of the data collected.

2.9 Measurement Issues

Measurement Error (Errors in Variables)

Obtaining accurate measurements of economic variables from data poses a substantial challenge (Griliches 1986). Data obtained from surveys, landings files, or secondary statistics typically contain measurement error, also called errors in variables. Measurement errors also arise in value measures of the capital stock or its services in a fishery where the purchase price of an individual vessel can differ widely from its current market value.

Proxy Variables

Errors can arise when using proxy variables. In some instances, a variable does not have an observable counterpart. Human capital or skipper skill and fishing effort are prominent examples in fisheries economics. Unless an observable indicator is found for the variable, the model has to be treated in the framework of missing variables, which leads to omitted-variable bias.

One of the most common solutions in such situations is the use of a proxy variable. For example, Squires et al. (2002) used various characteristics of skippers, such as specified as fishing experience and formal years of education, to proxy the human capital and managerial ability of skipper in Malaysian gillnet fisheries. A widely used proxy variable in fisheries is the use of fishing time for the unobserved composite input fishing effort. The proxy, by definition, contains measurement errors, but a poor proxy is often better than omitting the unobservable explanatory variable entirely.

Issues Arising in Estimation of Incomes and Profits

Surveys of fishers in developing and developed countries alike at the level of vessels or households face a problem in estimating income and employment, because fishers are often self-employed. Self-employed fishers typically have no familiarity with economists' definitions of income or private profit (Deaton 1995). Direct questions about income or profitability can yield misleading answers, particularly for individuals whose personal or business transactions are not clearly separated. Instead, as Deaton (1995) observes, surveys need to ask detailed questions about business or fishing operations, about sales and purchases, about quantities and prices, about taxes and transfers, about multiple business or household activities, about transactions in kind, and about assets. A measure of income or economic profit can then be derived from this detailed information by imposing an accounting framework on each economic agent's activities. This accounting framework includes measures of asset depreciation, where the assets to be measured include vessels, engines, gear, and equipment.

When fisheries are regulated by some kind of transferable property right, such as individual transferable quotas, licenses, or individual transferable effort quotas, these assets have financial and economic value and affect financial and economic measures of cost and profitability. When these rights are assigned to individuals rather than vessels, the financial and economic values of the vessel and the right are not confounded and can be individually measured. When the rights are assigned to vessels, the values of the right and vessel are confounded, so that the vessel value includes the value of the right.

Measurement of autoconsumption, or the fraction of consumption that is produced by a fisher or by the household without going through a market, is an issue of particular importance to developing countries. In many artisan fishing communities, large proportions of the population consume fish and other food items that are produced by the household, or are obtained through barter with other households. The usual practice is to collect data on the physical quantities of food and other autoconsumed items, and then value these by an imputed price, such as the market price of fish of a similar species and grade.

Measurement of the economic opportunity cost of labor and the share system's valuation of labor are both difficult issues when collecting fisheries data. Fishers' remuneration from the share system can be either directly asked or imputed from data collected on costs and returns. Any subsequent econometric analysis using these shares as explanatory variables typically runs the risk of simultaneity bias because the share paid is a function of both revenue and costs, both of which are components of profit. In such models, an alternative measure is to use labor's opportunity cost, or the wage payable for crew in their next best employment alternative. For example, in the United States, measures of wages for different employment categories are available from both state and federal sources. As it is difficult to actually pinpoint the type of employment a fisher might seek outside of fishing, a weighted average of the alternative wages is perhaps the best exogenous measure of fishing labor costs.

2.10 Outliers and Influential Observations

Estimates of economic models can be unduly influenced by one or several observations. Thus, when collecting economic data, care should be given to the accuracy of the data and to documenting and understanding any observations that are unusual in comparison to others. When collecting economic data from sample surveys, retention of the surveys for later reference is important; often, only upon using the data does the analyst actually observe an unusual observation. Errors can readily be made when entering data, such as adding extra zeros to a number or misplacing a decimal point or entering a number from a neighboring field. Reference back to the original survey can often clear up many an unusual observation.

Problems can also arise when sampling from a small population if one or more members of the population are very different from the rest of the population; such observations are called outliers or influential observations. Computing summary statistics, especially minima and maxima, can help catch mistakes in data entry. In some instances, another good summary statistic for catching errors is proportions or shares, such as the share of fuel in total costs. A comparison of a fuel cost share for one vessel with the mean or median to vessels of a comparable size and amount of fishing time expended can also help indicate if the fuel cost data are accurately recorded and if the observation is influential or not.

It is also important to ascertain whether any outliers present in the data are representative of the underlying population. An approach particularly well suited to identifying outliers in fisheries has been developed by Fox et al. (2004) which uses index numbers to identify 'extreme' values. It can be used for a variety of tasks such as to identify outliers for enforcement purposes due to misreporting. This approach recognizes that outlying observations provide important information about the extremes of variation in the population.

2.11 Missing Observations and Incomplete Data

Data can be incomplete in many different ways (Griliches 1986). First, they can be subject to undercoverage, which relates to sample design and refers to the possibility that a certain fraction of the relevant population was excluded from the sample by design or accident. This situation can yield a nonrandom sample from the population, and can also present a serious problem to economic analysis. Second, data can also be subject to unit non-response, which relates to the refusal of a unit or individual to respond to a questionnaire or interview, or the inability of the interviewers to find it. Third, data can also be subject to item non-response, which is the term associated with the more standard notion of missing data, such as questions unanswered, items not filled in, and so forth.

Missing data resulting in a nonrandom sample from the population can pose a serious problem. For example, a survey may be conducted to evaluate the effect of a closed area or season on fishers' profits where the fishery is comprised of several gear types and areas. Economists evaluating the effect would then construct a cost-and-earnings data set, but would also be unable to analyze data from fishers omitted from the sample. This omission could be due to non-response of the surveyed fishers to a particular survey instrument, or due to difficulty in intercepting fishers of a gear type who are often away at sea. If the cost-and-earnings data were easier to obtain for fishers with higher profits, the sample would not be representative of the population and insights from the analysis might be flawed.

Missing data can also be viewed as an extreme form of measurement error. The remedy for missing data depends on the original reasons for its absence. When data on explanatory variables (x) are missing, simply omitting observations for which one or more of the explanatory variables are missing is acceptable if these observations can be viewed as occurring randomly. If not, their omission causes sample selection bias. However, even when potentially discarded observations occur randomly, a small sample size may argue against ignoring the problem.

An approach to the missing data problem with explanatory variables is to replace the missing values with suitable proxy values and thereby allow the remaining information to improve estimation. However, replacing dependent variable (y) values with proxies is not recommended. Replacement of missing explanatory variables can be accomplished by predicting the missing value by using data from all other independent variables. This approach leaves the coefficient estimate of the variable with missing observations unaffected, but can improve estimation of the remaining coefficients because of the larger sample size, although it might also introduce some bias because of measurement error. A sensitivity analysis to this procedure using alternative forecasted x values can also indicate how robust the results are to the measurement error problem.

2.12 What Type of Data Should Be Collected?

The type of data that should be collected depends on the questions of interest and the nature of the analysis. One of the most fundamental considerations is whether cross-sectional, time series, pooled cross-sections of time series, or panel data are desired or even required.

Some questions can be answered by cross-sectional data, obtained via surveys. In developing countries, where the data are often collected by surveys, cross-sectional data and the use of recall information is often the only way to collect data. In developed countries, landings and revenue information are often available because governments routinely collect data at the point of land, but cost data are not. Cost data, even for bioeconomic or more dynamic models, are typically obtained by some form of cross-sectional survey and then applied to multiple time periods.

Types of Questions to Ask

The types of questions to ask when collecting economic data depend upon a number of factors, the most important of which include the intent of the analysis, budget limit, time available, and type of audience. Nonetheless, some general observations can be made.

Cost data are important for measuring resource rents, firms' economic and financial profits, certain kinds of efficiency, and impacts of different conservation and management policies on vessels, processors, among others. Cost data can be classified by at least two criteria: whether variable (changes with output or harvest) or fixed, and economic versus financial.

Variable Inputs and Costs

For fishing vessels, variable costs pertain to the costs of variable inputs, which are inputs that can be varied by the firm (vessel, processor) over a short time period. Cost data can be obtained either directly, or else indirectly, by collecting data on input usage and multiplying by an input price. For example, fuel and oil costs can be obtained directly as total expenditure on fuel and oil or indirectly as fuel and oil consumption multiplied by fuel and oil prices. Alternatively, fuel consumption can be calculated by dividing fuel expenditure by fuel price. As another example, the economic opportunity cost of labor can be obtained by collecting data on crew sizes and wage rates for alternative occupations of labor in different regions or ports and multiplying these together.

Variable inputs and their costs at a minimum include fuel and oil consumption and their costs or prices, labor costs and crew numbers (including captain), and in some instances, bait and/or ice. Labor (including captain, especially if not an owner-operator) typically receives a share of the 'gross stock', where the formula for calculating the crew and captain's share typically varies by fishery, gear type, region or port, and even vessel. When collecting labor shares, experience shows that a direct question on actual shares paid along with the crew share formulae are both useful. The direct question gives the desired information, and the crew share formulae allow double-checking or imputing values when the respondents do not answer the direct question on shares.

Gear and equipment are often variable costs, but are sometimes fixed costs, depending on how rapidly the gear and equipment depreciate and are replaced. If gear and equipment are replaced on an annual basis or some other relatively short period of time, then they can be considered variable costs, and their costs of purchase are the focus. If gear and equipment are longer lived – they are durable capital assets – then their treatment should be different.

Other variable inputs and their costs and/or prices also arise, including: bait; ice, chilling, or freezing; food (although food is sometimes part of crew share); at-sea trans-shipment costs; and hired captain. When an economic opportunity cost of labor is calculated, food cost is implicitly captured by this method and food cost would not explicitly enter into calculations of economic profit.

Some of the variable inputs can appear on a fishing firm's income statement only in terms of the value of the input, without any measure of quantity or price. Morrison (1993) observes that change in value, but without quantity units, requires the use of relevant price indexes or deflators, i.e., dividing the cost of an input or group of inputs by a relevant price index yields the quantity. The choice of which price index to serve as the representative unit price over time depends on the situation. The overall GDP implicit price deflator can be used when appropriate individual price indexes are unavailable or deemed unsuitable. In other instances, price indexes for specific industries or categories of inputs may be available, such as price indexes for fuel or other variable inputs.

Price indexes can be used with cross-sectional data to form time series for some variable inputs. For example, a cross-sectional survey of fuel costs yields a benchmark fuel cost for a given time period. This one observation on fuel cost or its price can be extended forward and backwards in time by assuming that fuel cost, or its price, varies over time at the same rate as the corresponding fuel price index. For example, if the fuel cost was $3,250 in 2000, but the 2001 measure was unavailable, and a fuel price index indicated a 5 per cent increase in 2001 over 2000, then the fuel cost could be computed as $3,250 × 1.05 = $3,412.50. This same approach can be used to fill in missing observations in time series and panel data.

Variable Costs and Inputs: Economic Opportunity Cost of Labor

In fisheries, the share system of remuneration to labor often does not provide a conventional measure of the wage. In this case an opportunity cost of labor may be used. When developing an opportunity cost for fishing labor, a crewmember whose skills are presumably specialized to fishing, is more likely to be easily replaceable by a workman of similar skills than a captain. Thus the wage or price of labor supplied by crewmembers would ideally be measured by the appropriate wage in the labor market. A captain is more likely to have skills that are readily transferable to occupations outside of fishing because skippers are more likely to have managerial, entrepreneurial, and perhaps have better mechanical skills than most crewmembers. Captains are also less replaceable than crewmembers. For these reasons, the opportunity cost assigned to captains is often assigned as a reasonable multiple of an ordinary crewmember (Squires 1984, 1992).

Unemployment benefits are also sometimes incorporated into the opportunity cost of labor to account for labor market distortions. This is appropriate when labor markets are not always fully competitive, in part because of a tendency for immobility of fishers contributing to unemployment. Dupont (1991), for example, incorporated unemployment benefits into her opportunity cost of fishing labor. The opportunity cost of labor then must reflect not only the next best employment alternative and the value of leisure time, but also the value of unemployment benefits.

Financial Fixed Costs

Financial or private fixed costs are those costs incurred by fishers that do not vary over a short time period (in contrast to the costs of a trip), and represent the market costs of inputs to production actually faced by firms or vessels. These financial fixed costs can include, but are not limited to: annual costs of total maintenance (across all gears); new safety equipment and its annual maintenance; new fishing gear; insurance (hull and liability, hull only, liability only, hull, liability, and health combined); electronics; depreciation; interest; professional and accounting; mooring or slip fees; engine maintenance; engine replacement; other maintenance; drydock and frequency; association and non-governmental organization membership costs; and other costs (e.g., travel, pingers, paint, trailer, storage and office, etc.).

Measures of depreciation are also needed to measure capital stocks and income flows from firms that use capital. A measure of depreciation is needed to obtain flows of income or profits because the flow of capital costs must be subtracted from revenues to obtain income or profits. Financial depreciation of an asset is an accounting convention, referring to a dollar set aside to allow for capital consumption allowances, which may or may not correspond to the actual use of economic resources and hence economic costs.

Depreciation of plant and capital equipment, as a financial cost, can be directly gathered by a question on a survey. Depreciation, as a financial cost, can also be calculated using historical book value, or the straight-line method where an estimated scrap value for the asset is subtracted from the original cost and the difference is then divided by the number of years of estimated life. Estimated economic life, however, can be difficult to determine and may require consultation with fishers or their organizations and retailers. When estimating the economic life of vessels recognition should also be given to the possibility of reconstructing and refurbishing vessels.

Dry-docking costs are appropriately measured as part of depreciation costs, as they represent a cost of avoiding depreciation that would otherwise occur. When periods in dry dock exceed a cycle greater than one year, costs can be apportioned over the number of pertinent years.

Measuring the quantities of individual or groups of capital such as various forms of gear and equipment is difficult. This is because firms' income statements or cost-and-earnings surveys may only provide cost rather than quantity data. In this case, costs for a combined group of individual pieces of gear and equipment can be divided by an appropriate price index, such as a producer price index for the most closely corresponding piece of capital, to yield a quantity index (see chapter 5). This approach overlooks important conceptual issues that arise when constructing measures of capital, such as physical depreciation of the stock and opportunity costs, capital gains or losses, and taxes, but it does provide an approximation of capital services.

The use of survey data allows more comprehensive measurement of gear and equipment. Survey data may yield costs and numbers of equipment, and even the year when the equipment was purchased, and the expected economic life. In these instances, the same approach of dividing costs by a producer price index for the group can be used. A more detailed approach is to create individual capital services prices for each category of gear or equipment and divide the corresponding costs by these prices to obtain quantity measures. Alternatively, each of the capital services prices can be weighted by its share of total costs to form a price for an aggregate of capital goods. The total cost of the collection of goods can then be divided by this composite price to obtain a corresponding quantity.

Determining an individual vessel's market value is another difficult data collection task because second-hand vessels are typically not sold on a regular basis. In addition, the historical or book value of individual vessels collected from regular survey data is often outdated and not representative of current market values. A potential source of vessel market values is the current insured value, where the notion holds that fishers' insured replacement value should be representative of current market values.

A measure of a representative vessel's value is usually less difficult to obtain than a specific fisher's vessel where there are sufficient number of vessels sold in a market. An alternative is to ask groups of fishers the values of representative vessels. Another alternative is to apply the 'Delphi' approach, whereby knowledgeable individuals, such as fishers, are queried for the market values of vessels, the responses collected and summary statistics calculated for these responses, these summary statistics (mean, minimum and maximum values) reported back to the individuals, and their second responses then summarized. Offer prices, which can differ from actual market values, can be obtained from advertisements in trade magazines or vessel brokers. While biased, offer prices may be preferred to seriously outdated market values.

Determining a vessel's value in developing countries, especially for artisan fisheries, is an even more daunting task, but all the above methods are applicable. The difficulty arises, however, when vessels are self-constructed and an active vessel market does not exist. In this instance, the respondent's best estimate of capital values and services may be the only way to obtain such data.

Economic Depreciation

In an economic, rather than a financial sense, depreciation is the decline in the value of an asset as it progresses from one vintage to the next oldest (Diewert, in press). In economics, the most commonly used approach to compute economic depreciation is the perpetual inventory method. In this approach capital services never actually reach zero, such that every unit of investment is perpetually part of the inventory of capital. The method essentially assumes the following:

$$K_t = (1 - \delta_t)K_{t-1} + I_{t-1} \tag{1}$$

where K_t is the capital stock at the beginning of time t, I_{t-1} is investment in period t-1 where the depreciation rate (δ_t) is usually defined as a constant, i.e., $\delta_t = \delta$ for all t (Morrison 1993).

Economic Fixed Costs

Economic fixed costs are those costs that do not vary with output and are valued at their opportunity cost, rather than what they might cost in a financial account. In placing an economic value on the existing stock of capital (or other fixed inputs), the most important issue is the existence and value of alternative uses. The sale value of the capital stock for a use other than the existing fishery provides a measure of the opportunity cost in terms of a permanent transfer. The foregone net earnings from an alternative use also provide a measure of the private cost of the temporary or seasonal use of that capital in the existing fishery.

In fisheries the economic cost of capital, especially the vessel and engine, is difficult to determine because of the absence of active markets, in which vessel prices are clearly established separately from licenses, the high degree of non-malleability of the capital stock because of very limited alternative uses, and the disparity between book or historical value and current market value. Over a one or two year time horizon the cost of the existing capital might be treated as a fixed cost with a low opportunity cost. However, over a much longer time period the cost of the existing capital stock is not fixed and the opportunity cost of either replacement or additional capital should be its market price.

Capital Services and Flows

Stocks of capital, such as the fishing vessel, yield a 'flow' of capital services (Hulten 1990). In some studies capital flows are assumed to be proportional to the size of the stocks, but the flow of services depends on how much the capital stock is used. For example, some vessels make more frequent and/or longer trips than others and, hence, are utilized more and provide greater flows of capital services. To account for these differences capital services can be measured as the estimated capital stock multiplied by an estimate of capital utilization. In fisheries, this is typically done by multiplying the measure of the capital stock by the most comprehensive measure of fishing time available, such as number of trips or days fished.

Capital Services Prices

Accounting for the economic contribution of capital to production is more difficult than accounting for the contributions of variable inputs such as labor or fuel. When a capital input is purchased for use at the beginning of an accounting period, the entire purchase cost cannot simply be charged to the period of purchase. This is because the benefits of using the capital asset extend over more than one period so that the initial purchase cost must be distributed over the useful life of the asset.

Economic theory specifies the service flow from the stock of capital as the appropriate measure of capital used. This, in turn, requires the construction of a corresponding data series that measures the service flow price (Morrison 1993). This concept is the basis for the notion of the user cost of capital, also called the rental or capital services price. This price depends on interest rate, tax structure, and depreciation and will vary for each component of the capital stock.

Joint and Common Costs and their Allocation

Joint costs are production costs incurred by the firm when two or more outputs are jointly produced. Joint costs and joint production costs can arise from an interdependent production process, a common input, or a common resource stock. Joint costs occur because a fisher finds it less expensive to incur costs relating to two or more activities than to incur costs individually for each.

Common costs occur when multiple products are produced together and arise if the costs for two (or more) outputs contain a fixed element common to both. Common costs apply to a setting in which production costs are defined on a single intermediate product or service that is used by two or more users.

Joint and common cost allocation is concerned with the distribution of these costs. An example of a joint cost that could be allocated among different fish species is the fixed cost of a vessel if it is used in fisheries that are seasonally and geographically distinct, or with different gear. If the operating costs in the various fisheries involved were nonjoint, they could also be directly allocated without difficulty. For example, the variable cost of fuel for swordfish-shark drift gillnet fishing is independent of fuel used when fishing for albacore with surface hook-and-line gear. Similarly, costs associated with driftnet gear are independent of surface hook-and-line gear and hence can be allocated relatively easily.

Measures of Capital Stock

The capital stock in a fishery is heterogeneous, that is, there are many different types of capital goods employed in the harvesting and processing of fish. The stock of harvesting capital includes the vessel hull, main, and perhaps auxiliary, engines, winches, booms, holds, chilling or cooling or freezing equipment, vessel electronics, and so on. The variety of capital assets often leads to treating some or most of the capital stock as homogenous even if they are not. For example, different types of capital or even the entire capital stock are often lumped together.

Market data are usually inadequate for the task of estimating the stock of capital and the price and quantity of capital services, which has led to the development of indirect procedures for inferring the quantity of capital, or the acceptance of flawed measures, such as book value (Hulten 1990). If there were an active rental market for capital services used and paid for in any given year, transactions in the rental market would give the price and quantity of capital in each time period. However, most of the capital stock is utilized by its owner and the transfer of capital services between owner and user only results in an implicit rent rather than an explicit payment observed in a market.

Practical options for directly measuring capital stocks are to find a direct estimate of the capital stock, to adjust book values for inflation, mergers, and accounting procedures, or to use the perpetual inventory method (Hulten 1990). One or more vessel attributes can also be used as proxy variables to measure the physical capital stock of a vessel where length, gross or net tonnage, or carrying capacity are the most widely used vessel attributes. The use of one or more vessel attributes is perhaps the most widely used measure of the capital stock in fisheries economics, in part because of the difficulty in obtaining accurate capital value information. Unfortunately, proxy variables, such as length or gross tonnage for the stock of capital, do not account for the different vintages (ages) of capital that embody different levels of technology, or the amount by which a stock of capital is actually used in a given time period.

2.13 Conclusions

In this chapter we have reviewed the various challenges and pitfalls of collecting and utilizing data for economic analysis of commercial fisheries. A major difficulty is that the economic data is often obtained from sources where the data were collected for other purposes. This creates potential biases in the data and the results that come from its analysis, but they can be mitigated in several different ways. The economic data that is used can be applied in efficiency studies, for measuring capacity and for estimating productivity, and also in bioeconomic models. These methods of analysis, which are discussed in subsequent chapters, provide key tools to improve fisheries management.

References

Arellano, M. and Honoré, B., 'Panel Data Models: Some Recent Developments', in J. Heckman and E. Leamer (eds), *Handbook of Econometrics* (vol. 5, North-Holland, 2001).

Baltagi, B., *Econometric Analysis of Panel Data* (New York: Wiley, 1995).

Deaton, A., 'Panel Data from Time Series of Cross Sections', *Journal of Econometrics*, 30 (1985): 109-26.

Deaton, A., *Understanding Consumption* (Clarendon Lectures in Economics, Oxford: Oxford University Press, 1992).

Deaton, A., 'Data and Econometric Tools for Development Analysis', in H. Chenery and T.N. Srinivasn (eds), *Handbook of Development Economics* (vol. 3A, North-Holland, 1995).

Deaton, A., *The Analysis of Household Surveys: A Microeconometric Approach to Development Policy* (Published for the World Bank, Baltimore: The Johns Hopkins University Press, 1997).

Diewert, W.E., *Issues in the Measurement of Capital Services, Depreciation, Asset Price Changes and Interest Rates* (National Bureau of Economic Research, in press).

Dupont, D., 'Testing for Input Substitution in a Fishery', *American Journal of Agricultural Economics*, 73/1 (1991): 155-64.

Fox, K.J., Hill, R.J. and Diewert, W.E., 'Identifying Outliers in Multi-Output Models', *Journal of Productivity Analysis*, 22 (2004): 73-94.

Grafton, R.Q., Nelson, H.W., and Turris, B., How to Resolve the Class II Common Property Problem? The Case of British Columbia's Multi-species Groundfish Trawl Fishery (Paper presented at the Conference on Fisheries Economics and Management in Honour of Professor Gordon R. Munro, Vancouver, Canada August 5 and 6 2004). Available at http://www.econ.ubc.ca/munro/472grnet.pdf.

Grafton, R.Q., Squires, D., and Fox, K.J., 'Private Property and Economic Efficiency: A Study of a Common-Pool Resource', *The Journal of Law and Economics*, 43/2 (2000): 679-713.

Griliches, Z., 'Economic Data Issues', in Z. Griliches and M.G. Intriligator (eds), *Handbook of Econometrics* (vol. 3, North-Holland, 1986).

Hannesson, R., 'The Bioeconomic Production Function in Fisheries: A Theoretical and Empirical Analysis', *Canadian Journal of Fisheries and Aquatic Sciences*, 40/7 (1983): 968-82.

Hilborn, R. and Mangel, M., *The Ecological Detective Confronting Models with Data* (Princeton, New Jersey: Princeton University Press, 1997).

Hsiao, C., *Analysis of Panel Data* (Econometric Society Monograph, no. 11, Cambridge, UK: Cambridge University Press, 1986).

Hulten, C.R., 'The Measurement of Capital', in Ernst R. Berndt and Jack Triplett (eds), *Fifty Years of Economic Measurement* (National Bureau of Economic Research Studies in Income and Wealth, vol. 54, University of Chicago Press, 1990).

Johnston, J., *Econometric Methods* (New York: McGraw-Hill Book Company, 1972).

Joseph, J., 'The Tuna-Dolphin Controversy in the Eastern Tropical Pacific Ocean: Biological, Economic, and Political Impacts', *Ocean Development and International Law*, 25 (1994): 1-30.

Leamer, E., *Specification Searches: Ad Hoc Inference with Nonexperimental Data* (New York: Wiley, 1978).

Maddala, G.S., *Econometrics* (New York: McGraw-Hill Book Company, 1979).

Morrison, C.J., *A Microeconomic Approach to the Measurement of Economic Performance: Productivity Growth, Capacity Utilization, and Related Performance Indicators* (New York: Springer-Verlag, 1993).

Pollock, R.A. and Wales, T.J., 'Specification and Estimation of Nonseparable Two-Stage Technologies: The Leontief CES and the Cobb-Douglas CES', *Journal of Political Economy*, 95/2 (1987): 311-33.

Squires, D., 'Public Regulation and the Structure of Production in Multiproduct Industries: An Application to the New England Trawl Fishery', *RAND Journal of*

Economics, 18/3 (1984): 232-47.

Squires, D., 'Productivity Measurement in Common-Property Resource Industries: An Application to the Pacific Coast Trawl Fishery', *RAND Journal of Economics*, 23/2 (1992): 221-36.

Squires, D. and Kirkley, J., 'Production Quotas in Multiproduct Industries: An Application to Multi-Species Fisheries', *Journal of Environmental Economics and Management*, 21 (1991): 109-26.

Squires, D. and Kirkley, J., 'Skipper Skill and Panel Data in Fishing Industries', *Canadian Journal of Fisheries and Aquatic Sciences*, 56 (1999): 2011-18.

Squires, D., Grafton, R.Q., Alam, F., Omar, Ishak H., 'Technical Efficiency of the Malaysian Artisanal Gill Net Fishery', *Environment and Development Economics*, 8 (2002): 481-504.

Wooldridge, J.M., *Econometric Analysis of Cross Section and Panel Data* (Cambridge, Mass.: MIT Press, 2002).

Wooldridge, J.M., *Introductory Econometrics: A Modern Approach* (South-Western, 2003).

Chapter 3

Measurement and Analysis of Efficiency in Fisheries

...a prime motive in any commercial fishery is to obtain the maximum income...
Ray Beverton and Sidney Holt (1957, p. 446) in On the Dynamics of Exploited Fish
Populations.

3.1 Introduction

Efficiency in fisheries is, in general, about doing the best we possibly can with the resources (fish stock and fishing inputs available) we have available. Improvements in efficiency by vessels are desirable provided that there exists a management structure that prevents biological and economic overexploitation. If not, increased efficiency, or the ability to catch more fish for a given amount of fishing effort, can be detrimental to sustainability.

Changes in efficiency of vessels are also strongly influenced by regulations. For example, imposing restrictions on what gear can be used by fishers affects the ability of vessels to harvest fish, and thus their efficiency. Consequently, a discussion about efficiency in a fisheries context is not possible without relating it to governance and management.

In this chapter we describe the relationship between fisheries governance and efficiency. We also provide explanations of different types of efficiency, and show how they can be measured and analyzed in a fisheries context using bioeconomic data. We conclude with applications of efficiency analysis in fisheries and what they imply about fisheries management.

3.2 Efficiency and the Common Pool Problem

In the first chapter we presented an overview of the economics of fishing. The principal insight is that, in the absence of any effective controls or management, a fishery will converge to a bionomic equilibrium where there is no economic surplus. In many fisheries, this bionomic equilibrium coincides with a lower fish stock than that which maximizes the sustained yield and will always be at a level of fishing effort that exceeds that which maximizes the economic surplus or economic profit from fishing. This is illustrated in Figure 3.1 where the sustained total revenue curve represents all the possible sustained yields for different levels of effort multiplied by a given price of fish. At zero fishing effort the sustained yield (and also sustained total revenue) is zero because no fishing takes place and, thus, is associated with high fish stocks. Where the total sustained revenue curve touches the horizontal axis at a high level of effort this represents the complete extinction of the fish stock. All points in between represent potential sustained revenues for given levels of fishing effort.

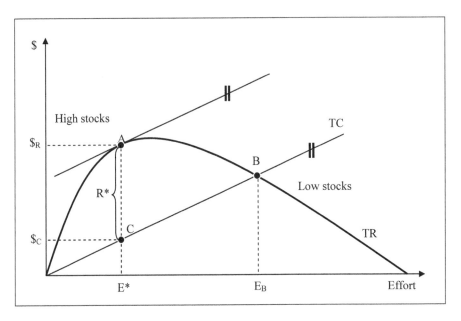

Figure 3.1 Technical and Scale Efficiency in a Fishery

The economic surplus from the fishery is represented by the difference between total revenue (TR), and the total costs of fishing (TC) that are assumed to be proportional to the level of fishing effort. The TC curve represents the minimum cost to catch fish for a given level of effort. The economic surplus is maximized at a level of fishing effort given by E* which coincides with the maximum distance between sustained TR (given by point A) and TC (given by point C) and generates a surplus per unit of effort of R*. By contrast, the bionomic equilibrium is given by point B, or where TR equals TC and coincides with a level of fishing effort equal to EB.

The bionomic equilibrium, and indeed any level of effort on the downward sloping side of the sustained revenue curve is technically inefficient because with fewer inputs (smaller level of fishing effort) it is possible to catch the same or greater quantity of fish (and thus have a higher sustained TR). Points between the origin and up until where the sustained revenue curve reaches a maximum are technically efficient because it is not possible to obtain a greater harvest (or TR) with a lower level of fishing effort. With the exception of E* all points from the origin to the maximum of the sustained revenue curve are technically efficient, but are scale inefficient in that they represent either too little (E < E*) or too much effort (E* < E) relative to the effort (E*) that maximizes the economic surplus. As a result, all levels of fishing effort, with the exception of E*, are overall inefficient. In other words, there is one and only one level of fishing effort that is both technically efficient and scale efficient, given by E*.

3.3 Allocative, Technical, Scale and Overall Efficiency

The notion of efficiency becomes more complicated in a dynamic setting where we value the present differently to how we value the future, and also where there are multiple inputs. In the case of multiple inputs, to achieve overall efficiency not only should vessels be technically efficient (use the minimum level of inputs to produce a given output) and scale efficient (produce at the optimal level of output), but they must also be allocatively efficient (use inputs in their cost-minimizing proportions). In most fisheries, regulations have been established to prevent harvesters from being economically efficient in an attempt to control fishing effort and over-fishing. This is most often accomplished by imposing input controls to limit the type of gear that can be used. However, such controls often prevent fishers using the cost-minimizing set of inputs, and restrictions on vessel size can prevent fishers from reaching their optimal scale of production. The consequence of such regulations is that the total cost of fishing per unit of effort increases over what it would be, thereby reducing the potential economic surplus. Ideally, a method of managing fisheries is required that allows us 'to have our cake and eat it too' whereby economic over-fishing is prevented and fishers have the opportunity to be as efficient as they can be.

Allocative Efficiency

Figure 3.2 illustrates the inefficiencies that input controls can generate for an individual fisher. It assumes that the fisher can use different combinations of two inputs (x_1 and x_2) to produce the identical level of output represented by the curve or isoquant (equal output curve) in the diagram. The fisher can either use a large amount of x_1 and small amount of x_2, or vice versa, and still produce the same level of output. The isoquant is curved to represent the idea that the two inputs cannot be perfectly substituted for each other. For example, it is virtually impossible to catch any fish even with all the fishing gear in the world without at least someone to operate it. The slope of the isoquant gets less steep as we use more of x_1 and less of x_2 to show that increasing the use of x_1 is subject to diminishing returns. The slope itself represents the ability to substitute between the two inputs while maintaining the same output. In turn, this depends on the extra or marginal output, also called the marginal product, from using more x_1 and the loss in marginal product from using less of x_2. This can be seen with the equation (3.1) where MP_i $i = 1, 2$ is the marginal product of input one and two, and Δx_i $i = 1, 2$ is the change in input one and two as we move down to the right (use more x_1 and less x_2) along the isoquant,

$$MP_1 \times \Delta x_1 + MP_2 \times \Delta x_2 = 0 \qquad (3.1)$$

where $\Delta x_1 > 0$ and $\Delta x_2 < 0$. In (3.1), the marginal product of input one multiplied by the change in input one (which is positive as we move down to the right along the isoquant) plus the marginal product of input two multiplied by the change in input two (which is negative as we move down to the right along the isoquant) must sum to zero for output to be unchanged. An equivalent way of writing (3.1) is as follows,

$$\frac{\Delta x_2}{\Delta x_1} = -\frac{MP_1}{MP_2} \qquad (3.2)$$

As the change in the two inputs gets infinitesimally small, the left hand side of (3.2) becomes the slope of the isoquant. From (3.2) we can see that the slope of the isoquant equals the negative of the ratio of the marginal products of the two inputs. The slope becomes flatter or less steep (the absolute value of the ratio of the marginal products gets smaller) as we move down to the right because of diminishing returns such that a successively greater amount of x_1 is required to offset the loss in output from using less x_2.

Assuming that all harvesting costs are variable, the total cost of production equals the amount of the two inputs used (x_1 and x_2) multiplied by their respective prices (p_1 and p_2), i.e.,

$$TC = (p_1 \times x_1) + (p_2 \times x_2) \qquad (3.3)$$

If input prices are fixed then the total cost of harvesting will only vary with the amount of the inputs used. It is also possible to have the same total cost with different combinations of the two inputs, as represented in Figure 3.2 by the straight line from the x_2 to the x_1 axis. Along this line, called the isocost (or equal cost) line, different combinations of the two inputs give the same total cost for given prices p_1 and p_2. The slope of the isocost line can be found by solving the equation for a line, or rewriting (3.3) so that x_2 is a function of the input prices (p_1 and p_2), the given total cost (TC) and x_1, i.e.,

$$x_2 = \frac{TC}{p_2} - \left(\frac{p_1}{p_2}\right) \times x_1 \qquad (3.4)$$

The first term on the right hand side of equation (3.4) is the vertical intercept, or the value of x_2 when $x_1 = 0$. The second term traces the movement down to the right along the isocost line for different values of x_1 until the isocost line touches the horizontal axis at the point TC / p_1. The slope of the isocost line is the negative of the ratio of the input prices given in brackets in (3.4). The higher is the price of input one relative to input two, the more steep will be the isocost line, reflecting the fact that fewer units of x_1 are required relative to units of x_2 to generate the same total cost.

An infinite number of isocost lines exist that represent different total costs. The further out is the isocost line from the origin, the larger the total cost. Thus to produce a given level of output at the minimum cost we would want to be on the isocost line that just touches, or is tangent to, the corresponding isoquant for that level of output. This is illustrated by point A in Figure 3.2. The cost minimizing combination of inputs that corresponds to point A is given by the quantities X_1^* and X_2^* which represents the allocatively (sometimes called the cost) efficient combination of inputs. This tangency point is where the ratio of the input prices equals the ratio of the marginal products of the inputs, i.e.,

$$\frac{p_1}{p_2} = \frac{MP_1}{MP_2} \tag{3.5}$$

If a regulator imposes input controls to restrict fishing effort, such as a maximum amount of x_1 that can be used by the fisher, denoted by MAX X_1 in Figure 3.2, then the fisher is no longer able to produce the output given by the isoquant at the lowest cost. If the fisher still tries to produce the level of output given by the isoquant she is now forced to use a different combination of inputs with more of the second input, and less of the first input due to the regulation imposed on x_1. This input combination is represented by point B in Figure 3.2 and is on a higher isocost than point A. The isocost curve with the regulated input x_1 is represented by the line D to B and corresponds to a greater level of total cost than the cost minimizing input combination because it is further from the origin. Consequently, the input combination given by point B is allocatively inefficient. The vertical distance from TC / p_2 to D represents the extra cost associated with the input control to achieve the level of output given by the isoquant. In other words, it represents the reduction in total costs if the fisher were able to be at the allocatively efficient input combination (point A) rather than at the inefficient combination (point B).

Typically, allocative efficiency is defined as the ratio of the isocost associated with the allocatively efficient input combination over the isocost of the actual input combination while preserving the actual input proportions. Adopting this approach makes the efficiency measure invariant to the units used to measure the inputs. The actual input proportions (ratio of input two over input one) at point B is the same as every point on a line, sometimes called a ray, drawn from the origin to point B.

Thus, in Figure 3.2, a measure of allocative efficiency is given by the ratio of the distance from the origin to point C (the isocost associated with cost-minimizing input combination) over the distance from the origin to point B (the isocost associated with actual input combination), i.e., OC/OB. A fishing vessel is allocatively efficient if this ratio is equal to 1.0, and is allocatively inefficient if the measure is less than 1.0. The smallest possible value is 0, which occurs when the optimal input combination is not to use any inputs and, thus, the total cost of production is zero.

So far, we have assumed that the cause of a fisher's allocative inefficiency is input controls, but fishers may still be inefficient because they have failed to optimize effectively. For example, a fisher may choose to employ more crew than she needs because she would rather have the extra help in case of an emergency rather than minimize costs. Alternatively, some fishers may inadvertently be at an inefficient input combination because they have given insufficient attention to their fishing operations.

Technical Efficiency

Technical efficiency is usually what fishery managers refer to when making efficiency comparisons across vessels, or over time. An input-orientated way of defining technical efficiency is the minimum amount of inputs required to produce a given level of output. In many fisheries, fishing vessels are not technically efficient because they use too many inputs, or are overcapitalized in the sense that a lower level of input (often measured in number of vessels) could be used to catch the same total harvest. Technical inefficiency may surface for many reasons, but a major cause is inputs controls that fail to prevent effort creep due to input substitution.

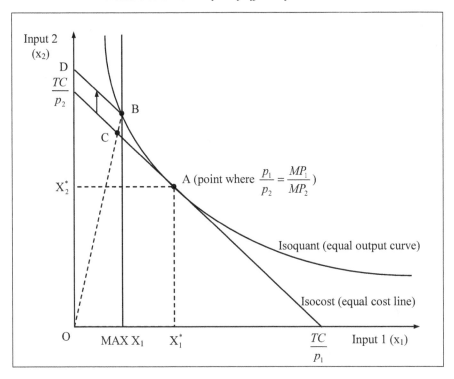

Figure 3.2 Allocative Efficiency and Input Controls

To better understand technical inefficiency, and how it differs to allocative inefficiency, we can use a similar diagram to that presented in Figure 3.2. In Figure 3.3, any input combination along the isoquant that represents 10 units of output is technically efficient because it represents the least amount of the two inputs, in different proportions, that are required to produce the same output. Thus point A, and indeed any point along the isoquant, is technically efficient. By comparing Figure 3.2 and 3.3 we can show that it is possible to be technically efficient (be on the isoquant in both figures), but be allocatively inefficient. For instance, point B in Figure 3.2 is technically efficient as it is on the isoquant, but is allocatively inefficient because it does not produce this output at the lowest cost input combination.

If we imagine that there also exists an input combination that produces the same output of 10 units given by the isoquant in Figure 3.3, but with greater amounts of the two inputs, then production at this point is technically inefficient. In other words, if it is possible to reduce the level of inputs and leave output unchanged then fishing must be technically inefficient. This is represented in Figure 3.3 by point B that corresponds to the same level of output of 10 units as at point A (or any point on the isoquant). Consequently, the distance A to B represents the reduction in inputs by the same proportion that would allow the fisher to produce the same level of output (10 units) and be technically efficient. Typically, the input-orientated measure of technical efficiency is defined as the ratio of the minimum inputs required to reach a given level of output over the actual inputs, while preserving the actual input proportions. As in Figure 3.2, the actual input proportions can be preserved by drawing a line, sometimes called a ray, from the origin to the actual input combination given at point B. The movement from B to A is called a radial contraction that allows the vessel to produce the same output, but with a smaller amount of inputs while maintaining the same input ratio (x_2 / x_1) at the actual production point B. Thus in Figure 3.3 a measure of input-orientated technical efficiency is given by the ratio of the distance from the origin to point A (the technically efficient point) over the distance from the origin to the actual production point B, i.e., OA/OB. If the fishing vessel is technically efficient then the value of the ratio is 1.0, and if it is technically inefficient it has a value of less than 1.0 with a lower bound of 0.

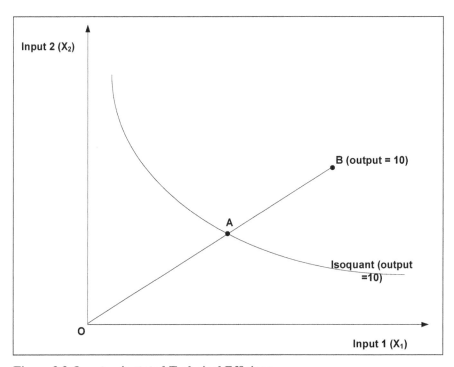

Figure 3.3 Input-orientated Technical Efficiency

Technical efficiency can also be represented with an output orientation, which is identical to an input orientation if there is a fixed relationship between the amount of inputs used and the level of harvest. With an output orientation, a fisher is technically efficient if she can produce the maximum output from a given set of inputs. In other words, in the output orientation we ask the question, can we produce the maximum output given our existing input combination? By contrast, with the input orientation we ask the question, can we produce the same amount of output, but with reduced inputs?

Both the input and output orientation for technical inefficiency are shown in Figure 3.4 that plots the relationship between the amount of fishing input or effort of a single vessel against its output or harvest of fish. Such a relationship is called a production function and represents the maximum output that can be produced for different levels of input. As the production function maps out 'best practice' or the boundary of the input-output relationship, it is also called production frontier.[1]

We would expect that for a single vessel that has no discernible impact on the total harvest or stock size that output should increase the more input we use. The production frontier in Figure 3.4 illustrates such a relationship, but also shows that the input-output relationship is subject to diminishing returns such that each extra unit of input gives less extra or marginal output then the previous unit of input. For example, for a skipper-owner of a vessel who hires a second crew member, the proportional increase in output would be less than that obtained from hiring the first crew member, and would get successively smaller with each extra crew hired. In reality, diminishing returns may not arise immediately, and with small amounts of an input it is even possible over some interval to have increasing returns such that the marginal output from an extra unit of the input increases with each successive unit of the input.

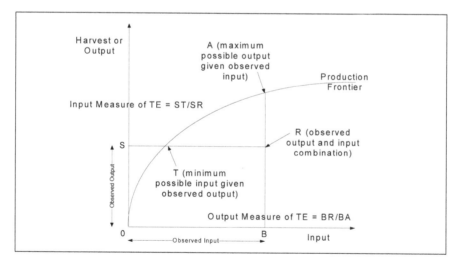

Figure 3.4 Output and Input Orientated Measures of Technical Efficiency

1   The common technology associated with multiple outputs can be represented by a *distance function* which is analogous to a single-output production frontier.

In Figure 3.4 any point on the production frontier is technically efficient because we are producing maximum output from a given input. Obviously, the more input we use the greater the output or harvest, but at each level of input we are producing the maximum output possible. Any point beneath the production frontier must be technically inefficient as it is possible to increase output without changing the amount of input used. Thus point A is technically efficient as it is on the production frontier, but point R is technically inefficient as it is beneath the frontier. The degree or level of technical inefficiency is given by the distance R to A — the larger the distance the greater the gap between actual and maximum output and, hence, the greater the technical inefficiency. For comparability across vessels, output-orientated technical efficiency is measured as a ratio of the distance BR (actual output produced with input level B) over BA (output produced with input level B if fisher were technically efficient), i.e., BR/BA. Thus if actual production is at the frontier, the measure is equal to 1.0 and represents the highest possible level of technical efficiency, but if production takes place beneath the frontier its value is less than 1.0, with its lowest possible value being 0.

In Figure 3.4 we can also represent the input orientated measure of technical inefficiency defined as the minimum level of inputs to produce a given output. Point T is on the production frontier and is, therefore, technically efficient. The minimum amount of input required to produce the output quantity S, consistent with frontier point T and actual production point R, is the distance S to T. However, if the actual production point is at point R then it implies that if the fisher were technically efficient she could reduce the amount of input given by the distance T to R, be at the frontier, and still produce the same output given by the vertical distance from the origin to S. Thus the larger is the distance T to R, the greater is the input orientated technical inefficiency.

Typically, the input-orientated measure of technical efficiency is given as a ratio defined by ST (minimum input level to reach the production frontier and produce the given output) over SR (the actual input level to produce the given output), i.e., ST/SR. If actual production is at the frontier the measure is equal to 1.0 and represents the highest possible level of technical efficiency, but if production takes place beneath the frontier its value is less than 1.0 with its lowest possible value being 0.

The input and output orientated measures of technical efficiency are identical if the production frontier is a straight line such that there is a fixed relationship or constant returns between the amount of input used and the output it generates. In the case of diminishing returns the output-orientated measure of technical efficiency will give a higher value than the input-orientated measure.

Technical Efficiency Comparisons

To understand how technical efficiency comparisons can be made across fishing vessels we can specify the following common production technology,

$$y_i = f(x_i) \times TE_i \tag{3.6}$$

The i subscript denotes observation or vessel i where $i = 1, 2, ..., N$. The variable y_i is the observed or actual output or harvest, x_i is the observed input used and TE_i is the technical efficiency of vessel i. All vessels are assumed to share the same production technology given by $f(\cdot)$ that efficiently transforms the effort expended in terms of input into units of output.

In the representation given by (3.6), differences in output across vessels are explained by variations in, one, the amount of input used and, two, differences in technical efficiency. Using the output orientated measure of technical efficiency as the ratio of observed output (y_i) to maximum possible (frontier) output ($f(x_i)$), we can obtain a measure of technical efficiency for each vessel that is bounded by 0 and 1.0, i.e.,

$$TE_i = \frac{y_i}{f(x_i)} \tag{3.7}$$

In this way, we can compare technical efficiency across vessels or groups (small or large, inshore or offshore, etc.) to ascertain what factors may be constraining efficiency among the fishing fleet. This can be illustrated in Table 3.1 using hypothetical data for three vessels where the common technology is defined by the function $f(x_i) = \sqrt{x_i}$, and is used to calculate the frontier output for all possible values of the input. The greater is the ratio of observed to frontier (maximum possible) output, the higher is the level of technical efficiency.

Table 3.1 Calculating Technical Efficiency with Hypothetical Data

Vessel	Observed Input	Observed Output	Frontier Output	Technical Efficiency
1	3	1.5	$\sqrt{3}$	$\dfrac{1.5}{\sqrt{3}} = 0.87$
2	4	1.0	$\sqrt{4} = 2.0$	$\dfrac{1.0}{2.0} = 0.5$
3	1	0.8	$\sqrt{1.0} = 1.0$	$\dfrac{0.8}{1.0} = 0.8$

Economic Scale and Overall Efficiency

Measures of allocative and technical efficiency represent different ways that fishing vessels are operating sub-optimally, that is, not doing the best they can with the inputs they have available. The concept of economic efficiency combines both allocative and technical efficiency, and is defined as the product of allocative efficiency and technical efficiency, i.e., Economic Efficiency = Allocative Efficiency × Technical Efficiency. In other words, a fisher must be both allocatively and technically efficient to be economically efficient for a given output level. As both measures of allocative and technical efficiency are defined as ratios between 0 and 1.0 then economic efficiency is also bounded between 0 and 1.0, where a value of 1.0 indicates a vessel is economically efficient. For the fisher to be overall efficient, she must also be at the optimal scale or level of production that maximizes economic profits, and be scale efficient. Thus overall efficiency requires both economic efficiency (which demands allocative and technical efficiency) and scale efficiency.

3.4 Predicting Efficiency

We have so far assumed that the production frontier is known and we can simply compare actual output or input combinations to optimal or frontier values and determine the efficiency of a fishing vessel. In reality, we first have to estimate or predict the efficient production frontier (and the associated isoquants) before we can calculate the efficiency of vessels. Two principal approaches to estimating a common and efficient production frontier are: one, data envelopment analysis (DEA), originally proposed by Farrell (1957), but first implemented by Charnes et al. (1978) and, two, stochastic frontier analysis (SFA) that was developed independently by Aigner et al. (1977) and Meeusen and van den Broeck (1977).

DEA uses mathematical programming to calculate a production frontier from the observed input-output observations across vessels. The major weakness of DEA in the fisheries context is that it assumes that all deviations below the common production frontier can solely be explained by technical inefficiency. In reality, a fisher may be beneath the frontier not because she is inefficient, but because of some stochastic or random event, such as bad weather. By contrast, SFA allows for the possibility of deviations above and below an estimated deterministic production frontier to arise from random events, other than technical efficiency. As a result, SFA is typically the preferred approach to efficiency analysis in fisheries.

Stochastic Frontier Analysis

The concept of a stochastic production frontier is very similar to the deterministic production frontier discussed previously except that it allows for the possibility of random events that shift the position of vessels in relation to the deterministic (non-stochastic) frontier. These random events are typically assumed to be both positive (such as a lucky fishing event) and negative (bad and unexpected weather). It is even possible that an especially large and positive realization of a random event may allow a vessel to be above the deterministic frontier for a particular point in time. This is illustrated in Figure 3.5 where the random event for a vessel i is denoted by v_i and is sufficiently positive to overcome the possible technical inefficiency of the vessel, and give it a higher output than the maximum possible without the random event. Equally possible is a negative realization of v_i that reduces stochastic output and would accentuate any possible technical inefficiency, and move the output even further below the deterministic frontier.

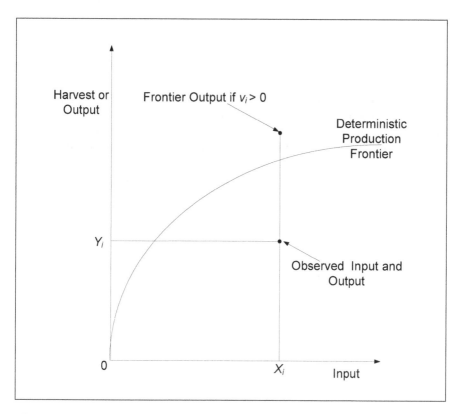

Figure 3.5 Deterministic Production Frontier and Stochastic Output

This representation of a stochastic frontier is similar to the deterministic frontier given in equation (3.6), but with the addition of this random error term, v_i, i.e.,

$$y_i = f(x_i; \beta) \times e^{v_i} \times TE_i \qquad (3.8)$$

The i subscript denotes observation or vessel i where $i = 1, 2, ..., N$, y_i is observed output, x_i is observed input and TEi is technical efficiency of vessel i. The common deterministic production frontier is given by $f(x_i; \beta)$ where the β represents the vector of parameters we need to estimate to construct this unknown deterministic frontier, while the stochastic production frontier is given by $f(x_i; \beta) \times e^{v_i}$ which includes the random error term, v_i. The random error term is given as an exponent of the number e (which equals 2.71828...). The number e has a very special property and allows us to rewrite the stochastic frontier in a much simpler, but equivalent, way.

Using equation (3.8), an individual measure of technical efficiency for vessel i is the ratio of observed output to maximum feasible output given the random error, i.e.,

$$TE_i = \frac{y_i}{f(x_i; \beta) \times e^{v_i}} \tag{3.9}$$

If we knew the individual random errors for each vessel we could estimate the deterministic production frontier and then calculate individual technical efficiencies. Unfortunately, the individual v_i term is not known and, instead what we have is a 'composed' error term (ε_i) that is the difference between observed output (y_i) and the predicted deterministic frontier output given by $f(x_i; \beta)$, i.e.,

$$\varepsilon_i = y_i - f(x_i; \beta) \tag{3.10}$$

The predicted and deterministic frontier output is obtained from estimating a production frontier using observed inputs and outputs for all vessels in a given sample. The inputs, given by x_i, are assumed to influence the deterministic production frontier, but are also assumed to be uncorrelated with either the random term v_i, or technical efficiency. In other words, the observed inputs determine the deterministic production frontier, but not the technical inefficiency or the random effects associated with fishing. Deviations from the actual output and the predicted deterministic frontier output occur because of, one, random events that can be positive or negative and given by v_i and two, technical inefficiency which is typically defined as u_i. As technical inefficiency always places us beneath the deterministic production frontier it must always reduce the composed error term ε_i given in (3.10). In other words, we can equivalently write the composed error term, which is observable given the predicted deterministic frontier output, as follows,

$$\varepsilon_i = v_i - u_i \tag{3.11}$$

where $u_i \geq 0$. The larger the value of u_i the greater the level of technical inefficiency because the further we are below the deterministic production frontier. Consequently, for a vessel to be on the deterministic frontier, and therefore technically efficient, it must be the case that $u_i = 0$.

As only ε_i is directly observable, but u_i and v_i are not, we must estimate or predict u_i using the composed error term and by placing restrictions on the nature of the distributions for u_i and v_i. Typically, analysts assume the u_i are distributed as a truncated normal distribution (although other distributions can also be assumed) and that the v_i terms are independently (values are unaffected by u_i and inputs) and normally distributed.[2] There are several software packages that allow us to calculate technical inefficiencies using SFA from observations of output and inputs. Some of these packages also allow us to use different distributions for u_i to test the sensitivity of our results to the distributional restrictions. When using such packages we must also undertake hypothesis tests in terms of the composed error term and, in particular, a test that there are no technical inefficiency effects. If there are no technical inefficiency effects then $u_i = 0$ for $i = 1, 2, ..., N$ and, therefore, $\varepsilon_i = v_i$ and the composed error term will be symmetric, and by assumption, normally distributed. By contrast, if there are technical inefficiency effects then the composed error term will be given by $\varepsilon_i = v_i - u_i$ and the distribution will be negatively skewed (mean is less than the mode and with a long thin tail to the left) and will no longer be symmetric. Thus a test for the existence of technical inefficiency effects is a test for negative skewness of the composed error term.

An alternative to making distributional assumptions about u_i is to estimate a production frontier using panel data where we have multiple observations over time of the same vessels (Berger 1993). In this 'distribution free' approach a stochastic frontier is estimated in each time period where the composed error includes random effects for each vessel for that time period (v_{it}) and also a time-invariant technical inefficiency for each vessel (u_i) that is independent of the inputs. Assuming that the random effects tend to zero over time such that the positive shocks more or less cancel out the negative shocks in the long run, then the average of the residuals over time is a predictor of technical inefficiency, i.e., $\hat{\varepsilon}_i = \left(\frac{1}{T}\right)\sum_{t=1}^{T}\hat{\varepsilon}_{it} \cong \bar{u}_i$ where T is the number of time periods. Thus for this assumption to hold true, and for the approach to be appropriate to estimate technical inefficiency, a sufficiently large enough T is required for the v_{it} to 'average' to zero. To implement the 'distribution free' approach it is also necessary to ensure that $u_i \geq 0$. This is done by transforming the predicted values of technical inefficiency, the $\hat{\varepsilon}_i$ as follows,

$$\hat{TE}_i = e^{-[\hat{\varepsilon}_i - \min_i(\hat{\varepsilon}_i)]} \qquad (3.12)$$

---

2   A truncated normal distribution requires us to estimate the mode (most frequent value in the distribution) of the distribution in addition to the individual u_i.

where the vessel with the lowest predicted technical inefficiency, denoted by $\min(\hat{\varepsilon}_i)$, is assumed to be on the production frontier, i.e., $\hat{TE}_i = 1.0$.

Predicting technical inefficiencies for vessels provides valuable information as to what vessel characteristics (types, home port, size, etc.) may be affecting efficiency. If we have data on observed inputs and outputs over multiple periods we can also test whether changes in fisheries management at a particular point in time influence technical efficiency. Determining what factors may be affecting technical efficiency, and their size and significance, is the major reason analysts use SFA in fisheries. This hypothesis testing (such as a test whether large vessels are more technically efficient than small vessels) is accomplished by estimating, in addition to the stochastic frontier, a technical inefficiency model of the following form,

$$u_i = g(z_i;\alpha) + \omega_i \qquad\qquad (3.13)$$

where u_i is estimated technical inefficiency for vessel i and z_i is a vector of individual vessel and environmental characteristics for vessel i that influences observed output indirectly through effects on technical inefficiency. In (3.13) the term α is a vector of parameters to be estimated using the sample data and typically would include an intercept term, while ω_i is an additional random error term that may be included depending on the estimation procedure.[3]

In current practice, the stochastic frontier model and the technical inefficiency model are combined in one estimation procedure which allows for the possibility of the z_i to be in the predicted production frontier, and also in the predicted technical inefficiency model. The choice of whether a variable belongs in the production frontier or the technical inefficiency model depends on our a priori understanding as to whether it affects the production frontier directly, or determines technical inefficiency, or both. For example, we should include some measure of fish stock size or abundance when estimating a stochastic frontier because it is likely to affect the production frontier. However, we would not expect technical inefficiency to depend on the size of the fish stock.

A final word on estimating, SFA is the choice of the functional form for both the production frontier and the technical inefficiency model. For tractability, many analysts assume a log-linear specification for the frontier and a linear specification for the technical inefficiency model. A log-linear model is simply a model that is linear in the natural logarithm of the inputs and so-called environmental variables. It requires that no variable is raised to a power other than one, and that no estimated parameter is multiplied by another parameter after taking the natural logarithm of both sides of the stochastic frontier specification. Thus if we take the natural logarithm of

---

3   See Kumbhakar and Knox Lovell (2000, pp. 262-77) for further details about how to estimate the technical inefficiency model.

the stochastic frontier specified by equation (3.8) we obtain the following log-linear specification,

$$\ln y_i = \ln f(x_i; \beta) + v_i - u_i \qquad (3.14)$$

As the common deterministic frontier is an arbitrary function that must be estimated, we can redefine (3.14), with no loss of generality, as

$$\ln y_i = f(\ln x_i; \beta) + v_i - u_i \qquad (3.15)$$

Typically, analysts assume that the deterministic production frontier can be approximated by a so-called flexible functional form which allows for 'flexibility' in how inputs can be substituted for each other while keeping output unchanged, and also flexibility as to whether the frontier exhibits constant or diminishing returns. The trade-off with flexible functional forms is that flexibility requires that more parameters be estimated, which may be difficult with only a small number of observations and if there are many inputs and environmental variables.

A widely used flexible functional form for the deterministic production frontier is the translog functional form. An advantage of the translog specification is that a much simpler functional form – the Cobb-Douglas – which is restrictive in terms of its assumptions regarding input substitution, can be tested to determine whether it provides an adequate representation of the data.

If there are S inputs (denoted by j) and M environmental variables (denoted by k) a Cobb-Douglas production frontier for $i = 1, 2, ..., N$ vessels is given by equation (3.16)

$$\ln y_i = \beta_0 + \sum_{j=1}^{S} \beta_j \ln x_{ji} + \sum_{k=1}^{M} \gamma_k \ln z_{ki} + v_i - u_i \qquad (3.16)$$

where β_0 (the intercept term), β_1, $\beta_2 ..., \beta_S$ and γ_1, $\gamma_2 ..., \gamma_M$ are parameters to be estimated using observations of the inputs and outputs of vessels and the environmental variables.

A technical inefficiency model linear in the environmental variables and that could be estimated simultaneously with (3.16) is given below,

$$u_i = \alpha_0 + \sum_{k=1}^{M} \alpha_k z_{k.} \qquad (3.17)$$

where α_0, α_1, ... α_M are parameters to be estimated.

Revenue, Cost and Profit Frontiers

All our analysis in terms of technical efficiency has been presented in terms of production frontiers, or the direct relationship between inputs and outputs. It is also possible to estimate frontiers where the dependent variable could be total harvesting revenue, total variable costs or profit. In these cases the resulting frontier indirectly provides information on the production frontier.

The choice of what frontier to estimate depends on data availability and the research question. For instance, if the analyst wishes to obtain estimates of allocative efficiency then either a cost frontier must be estimated directly, or it must be constructed from estimates of a production frontier. When a cost frontier is directly estimated the technical inefficiency term is added to the random error, i.e., $\varepsilon_i = v_i + u_i$. In other words, the inefficiency term defined by $u_i \geq 0$, increases costs and places vessels above the minimum cost frontier. The production frontier can also be obtained using estimates from the cost frontier, and vice versa, when the estimated function form of the frontier is said to be self-dual, such as the Cobb-Douglas form.

3.5 Fisheries Applications

There are many possible applications of efficiency analysis in fisheries. We present three examples: one, a test of the technical efficiency effects of input controls, two, an analysis of the influence of individual output controls on economic efficiency and, three, an examination of how vessel, gear and skipper and crew characteristics affect technical efficiency.

Technical Efficiency Effects of Input Controls

Using two different sources of individual vessel data (1990-1996 and 1994-2000), two separate stochastic production frontiers were estimated by Kompas et al. (2004) for the Northern Prawn Fishery (NPF) located off the north coast of Australia. The NPF is one of Australia's most valuable fisheries and has been regulated by input controls since 1977, when the total number of vessels in the fishery was fixed at 292, despite the fact less than 200 vessels were active in the NPF at that time.

High returns in the fishery in the 1970s and 1980s encouraged individuals with licensed, but inactive, vessels to participate in the fishery such that by 1981 the number of vessels fishing prawns was almost at its regulated limit of 292. In addition to an increase in the number of active vessels, the vessels have become more effective at catching prawns over time. To address this effort 'creep' controls were introduced in terms of engine power and hull size, designated as A-units, coupled with buybacks of vessels denominated in A-units.

Using 1990-1996 data Kompas et al. (2004) estimated a Cobb-Douglas stochastic production frontier given by (3.18),

$$\ln Y_{it} = \beta_0 + \beta_1 \ln \text{crew}_{it} + \beta_2 \ln \text{effort}_{it} + \beta_3 \ln \text{gear}_{it} + \beta_4 \ln \text{fuel}_{it} + \\ \beta_5 d90 + \beta_6 d92 + \beta_7 d94 + \beta_8 T + v_{it} - \mu_{it} \tag{3.18}$$

where Y_{it} is the output of banana prawns by observation i in period t, crew is number of persons per boat, including the skipper (or owner-operator), and effort is the average number of fishing days. Fuel represents all input expenditures (fuel, oil, and grease). Gear is the vessel's head-rope length which is a measure of trawling capacity and $d90$, $d92$ and $d94$ are year-dummies for 1990, 1992 and 1994 to account for weather related anomalies that represent years of abnormally low rainfall prior to harvest which reduces the fishing stock of prawns. As stock numbers are not available the time trend given by T captures non-specific effects on harvest.

To test for changes in technical efficiency over the period and whether controls on engine size and hull capacity (A-units) have had an influence, they also estimated the following technical inefficiency model.

$$\mu_{it} = \delta_0 + \delta_1 \text{A-units} + \delta_2 \text{gear} + \delta_3 \text{skipper} + \omega_{it} \tag{3.19}$$

where the ω_{it} is an error term to account for random differences in efficiency across vessels and skipper is a dummy variable which equals 1.0 where the captain is not an owner-operator.

They also tested whether the technical inefficiency effects are deterministic rather than stochastic, i.e., tested the following hypothesis,[4]

$$\gamma = \frac{\sigma_\mu^2}{(\sigma_v^2 + \sigma_\mu^2)} = 0 \tag{3.20}$$

where γ is an estimate of the relative importance of technical inefficiency effects in terms of the composed error term. If $\gamma = 0$ then there is no evidence of technical inefficiency effects. Given the estimated value of γ found by Kompas et al. (2004) exceeds 0.80, and is significant at the 1 per cent level, they found that the estimation of a stochastic frontier is appropriate. In a separate calculation, they also tested whether all technical inefficiency effects are absent, i.e.,

---

4   This involves a likelihood test where the null hypothesis is $\gamma = 0$ against the alternative $\gamma > 0$. Further details are available in Coelli et al. (1998).

$$\delta_0 = \delta_1 = \delta_2 = \delta_3 = 0 \qquad\qquad (3.21)$$

which was also rejected at the 1 per cent level of significance. As a result, Kompas et al. concluded it was appropriate to estimate (3.18) and (3.19). The results of their estimation are presented in Table 3.2.

Table 3.2 Parameter Estimates of the Stochastic Production Frontier and Technical Inefficiency Models in the Northern Prawn Fishery

Model	Coefficient
Stochastic production frontier	
Constant	6.42***
	(0.97)
Crew	-0.10
	(0.19)
Effort	0.38***
	(0.15)
Gear length	0.20*
	(0.11)
Fuel	0.27**
	(0.13)
Time trend	-0.05**
	(-0.02)
Year 1990	-0.62***
	(0.13)
Year 1992	-0.57***
	(0.10)
Year 1994	-0.62***
	(0.09)
Technical inefficiency model	
Constant	2.91
	(3.78)
A-unit	-1.52**
	(0.95)
Head rope length of gear	1.53**
	(0.85)
Skipper	0.73
	(0.92)
Mean Technical Efficiency	0.725

Notes: *, ** and *** denote statistical significance at the 0.10 level, 0.05 and 0.01 level respectively. Numbers in parentheses are asymptotic standard errors.

Source: Kompas et al. (2004).

Table 3.2 shows two important results of input controls in the Northern Prawn Fishery. First, controls on A-units (hull and engine size) by the regulator has had the net effect of reducing technical efficiency (or raising technical inefficiency) because the estimated coefficient in the technical inefficiency model is statistically significant and negative. In other words, for the average vessel an increase in A-units lowers technical inefficiency (raises technical efficiency). Second, Kompas et al. (2004) found that because of controls on A-units in the 1980s fishers have tended to substitute to increased headrope length so as to increase their fishing power. Unfortunately, the technical inefficiency model indicates that such input substitution has raised technical inefficiency (lowered technical efficiency) because its estimated coefficient is positive and statistically significant.

The efficiency analysis indicates that input controls on hull size and engine power and the substitution to unregulated inputs, such as headrope length, have reduced technical efficiency in the NPF. Such an outcome runs counter to the stated objective of the fishery regulator to both maximize economic efficiency and ensure the sustainability of the resource.

Economic Efficiency and Individual Output Controls

In 1979 the total number of vessels permitted to fish for halibut using longlines in British Columbia (BC) was limited to 435 vessels, although the number of active vessels at that time was just over 300. High returns in the fishery motivated fishers with the permitted halibut fishing licence to become active in the fishery. As a result, a decade later all halibut licences were fully utilized. The increase in the number of vessels also coincided with larger crew sizes and more sophisticated fishing gear such that the regulator consistently reduced the length of the fishing season to prevent the desired total harvest from being exceeded. Consequently, fishing season length declined from 65 days in 1980 to just 6 days a decade later, despite the fact that total catch was almost twice the level in 1990 than it was 10 years previously.

The prospect of a 1991 fishing season of just two days in 1991 and declining returns provided the impetus for fishers to support the introduction of individual vessel quotas (IVQs) in 1991. Initially the IVQs were introduced on a two-year trial basis and were allocated gratis to existing halibut licence holders based on their vessel length and their best catch over the previous four years. Following the two-year trial, IVQs were made transferable between halibut licence holders in 1993 although restrictions were placed on the amount of quota that could be used on any one vessel and also on their divisibility.

Using individual vessel data available from a sample of vessels in 1998, 1991 and 1994, Grafton et al. (2000) estimated a Cobb-Douglas stochastic production frontier given by (3.22).

$$\ln H_{it} = \beta_0 + \beta_1 \ln K_{it} + \beta_2 \ln L_{it} + \beta_3 \ln F_{it} + \beta_4 \ln B_{it} + \nu_{it} - u_{it} \quad (3.22)$$

where H is halibut catch of vessel *i* in period *t*, K is vessel hull length, *L* is number of crew multiplied by the number of weeks the vessel was halibut fishing, *F* is fuel consumption in liters and *B* is a measure the exploitable halibut biomass. The hypothesis test that $\gamma = 0$ was rejected at the 5 per cent level of significance which indicates that technical inefficiency effects exist in the data and account for more of the variability than random factors.

Using parameter estimates from (3.22) Grafton et al. (2000) also calculated a short-run cost frontier, self-dual to the production frontier, after first fixing vessel size equal to its observed length. In this way, they were able to obtain estimates of allocative efficiency as well as technical efficiency and also economic efficiency. Their estimates of the stochastic production frontier and mean technical efficiency are presented in Table 3.3.

Table 3.3 Stochastic Production Frontier for the British Columbia Halibut Fishery

Variable	Coefficient
Constant	2.2436
	(1.4287)
Vessel length	1.0294*
	(0.2221)
Labor	0.4122*
	(0.1175)
Fuel	0.2769
	(0.0781)
Biomass	1.0281*
	(0.3508)
Mean Technical Efficiency	0.56

Notes: * denotes statistical significance at the 0.05 level. Numbers in parentheses are asymptotic standard errors.

Source: Grafton et al. (2000).

Various tests of hypotheses were performed using the estimated measures of allocative, technical and economic efficiency for both small and large vessels. They found that there was initially a decline in some measures of efficiency from 1988 to 1991, primarily due to an almost 50 per cent decline in the total harvest and severe autumn storms in 1991. After transferability of IVQs was allowed in 1993, fishers were able to adjust their harvest to the appropriate scale of harvest and by this time had learnt to spread their fishing effort and adjust the mix of inputs in a better way to take advantage of the increase in the fishing season to 245 days from its 1990 level of 6 days. As a result, Grafton et al. (2000) find substantive evidence that both short-run technical and economic efficiency increased for small and large vessels between 1991 and 1994.[5]

Technical Efficiency and Vessels and Skipper Characteristics

In a study of the artisanal fishing communities of the west and east coasts of Malaysia, Squires et al. (2003) used individual vessel data to examine what factors may be constraining technical efficiency. Their study has immediate practical value because it provides information about what strategies can be used to assist fishing communities.

Squires et al. (2003) estimated a separate translog stochastic production frontier for the west and east coasts due to differences in resource availability and socio-economic conditions between the two locations. The function separately estimated on both coasts is given by (3.23).

$$\ln Y_i = \beta_0 + \beta_1 \ln K_i + \beta_2 \ln L_i + \beta_3 \ln T_i + \beta_4 \ln N_i + \beta_5 \ln OD_i +$$
$$\beta_6 \ln K_i^2 + \beta_7 \ln L_i^2 + \beta_8 \ln T_i^2 + \beta_9 \ln N_i^2 + \beta_{10} \ln K_i \ln L_i + \beta_{11} \ln K_i \ln T_i \quad (3.23)$$
$$+\beta_{12} \ln K_i \ln N_i + \beta_{13} \ln L_i \ln T_i + \beta_{14} \ln L_i \ln N_i + \beta_{15} \ln T_i \ln N_i + v_i - u_i$$

where total output (catch) in kilograms for each vessel i is denoted by Y_i and is the geometric mean of fifteen species of fish plus prawns (where revenue shares serve as weights). The inputs are specified as service flows by multiplying the stocks of capital and labor by days at sea. The vessel capital stock (K_i) is a volumetric measure given by vessel gross registered tons; labor (L_i) is the number of crew employed per vessel for the month, including the captain; and the gill net capital stock (N_i) is measured by its length in meters multiplied by the number of hauls of the gill net per day. The number of trips per month (T_i) represents variable input usage (e.g., diesel and/or gasoline, lubricant and/or oil, ice, container/polythene, and miscellaneous variable inputs). Distance from shore to the fishing ground is specified in nautical miles (OD_i) and is intended to capture environmental effects, providing for differences in resource conditions that vary by distance from shore and by water depth. The null hypothesis of no technical inefficiency effects $(\gamma = 0)$ was rejected at the 1 per cent level of significance. The mean levels of technical efficiency on the east and west coasts were 0.84 and 0.85 indicating an overall high degree of efficiency.

To test the factors that might be contributing to technical inefficiency they also estimated the following technical inefficiency model for east coast vessels,

---

5   See Grafton et al. (2000) footnote 61 and Table 8.

$$u_i = \delta_0 + \delta_1 EXLIH + \delta_2 EXLIE + \delta_3 EXLIN + \delta_4 FEXP \quad\quad (3.24)$$
$$+ \delta_5 MESH + \delta_6 FSIZE + \delta_7 D_{CH} + \delta_8 D_{CT} + \delta_9 D_{NOP} + \delta_{10} D_{SM} +$$
$$\delta_{11} D_P$$

where *EXLIH*, *EXLIE*, and *EXLIN* are the remaining economic life, in years, of the vessel hull, engine, and gill net as estimated by respondents; *FEXP* is years of fishing experience for the captain; *MESH* is mesh size in meters; and *FSIZE* is the family size of the captain. The five *D* terms are dummy variables and are equal to one when: the vessel has a Chinese captain (*CH*); the captain has participated in a Malaysian fisher training program (*CT*); the captain is not the owner of the vessel (*NOP*); the vessel is small (*SM*) — defined as less than 5 gross registered tons; and the captain has a primary education (*P*).

Table 3.4 Technical Inefficiency Model for East Coast Malaysia Gill Net Fishers

Variable	Coefficient	Standard Error
Constant	4.1338***	0.7067
EXLIH	-0.0261	0.0277
EXLIE	-0.0981**	0.0262
EXLIN	-0.0366	0.0479
FEXP	-0.0254*	0.011
MESH	0.4247	0.9968
FSIZE	-0.0766*	0.0363
D_{CH}	-0.4335	0.2264
D_{NOP}	-0.0847	0.2296
D_{CT}	-0.4763	0.4856
D_{SM}	-0.5088*	0.1754
D_P	-0.4887**	0.1511

Notes: *, ** and *** denote statistical significance at the 0.10 level, 0.05 and 0.01 level respectively. Numbers in parentheses are asymptotic standard errors.

Source: Squires et al. (2003).

The estimated coefficients of the technical inefficiency model using the east coast data from Squires et al. are presented in Table 3.4. The negative and statistically significant coefficient for *EXLIE*, *FEXP* and *FSIZE* results indicate that the longer the remaining life of the engine, the greater the fishing experience of the captain and larger the family size the greater the average technical efficiency (or the lower the technical inefficiency). The only significant coefficient for the dummy variables are the size of vessel (D_{SM}) and education of the captain (D_P) which are both negative. This implies that smaller vessels have higher average technical efficiency than larger vessels (lower technical inefficiency) and that vessels with captains that have a primary education also have a higher level of technical efficiency.

Overall, Squires et al. (2003) find high levels of technical efficiency, significant differences between the west and east coasts in terms of efficiency, and few benefits in terms of technical efficiency from improvements in gear and equipment. The implication of their study is that development assistance efforts should be directed away from the harvesting sector towards other priorities in artisanal fishing communities.

3.6 Conclusions

Efficiency analysis allows fishery managers to assess what factors affect economic performance, and to also assess the impacts of fisheries regulations. The primary tool for measuring efficiency in fisheries is stochastic frontier analysis that tries to account for the randomness of harvesting as well as individual vessel differences in performance. It requires individual vessel data and measures of outputs and inputs, revenues, costs or profits. The potential applications of stochastic frontier analysis in fisheries are limited only by the research questions managers have about the economic performance of harvesters, and the consequences of regulations.

References

Aigner, D.J., Lovell, C.A.K. and Schmidt, P., 'Formulation and Estimation of Stochastic Frontier Production Models', *Journal of Econometrics*, 6 (1977): 21-37.

Berger, A.N., 'Distribution Free' Estimates of Efficiency in the U.S. Banking Industry and Tests of the Standard Distributional Assumptions', *Journal of Productivity Analysis*, 4/3 (1993): 261-92.

Beverton, R.J.H. and Holt, S.J., On the Dynamics of Exploited Fish Populations (London: Her Majesty's Stationery Office, 1957).

Charnes, W., Cooper, W., and Rhodes, E., 'Measuring the Efficiency of Decision Making Units', *European Journal of Operational Research*, 2/6 (1978): 429-44.

Coelli, T., Prasada Rao, D.S. and Battese, G.E., *An Introduction to Efficiency and Productivity Analysis* (Boston: Kluwer Academic Publishers, 1998).

Farrell, M.J., 'The Measurement of Productive Efficiency', *Journal of the Royal Statistical Society*, 120 (1957): 253-90.

Grafton, R.Q., Squires, D. and Fox, K.J., 'Private Property and Economic Efficiency: A Study of a Common-Pool Resource', *Journal of Law and Economics*, 43 (2000): 679-713.

Kompas, T., Che, T.N. and Grafton, R.Q., 'Technical Efficiency Effects of Input Controls: Evidence from Australia's Banana Prawn Fishery', *Applied Economics*, 36 (2004): 1631-41.

Kumbhakar, S.C. and Knox Lovell, C.A., Stochastic Frontier Analysis (Cambridge, UK: Cambridge University Press, 2000).

Meeusen, W. and van den Broeck, J., 'Efficiency Estimation from Cobb-Douglas Production Functions with Composed Error', *International Economic Review*, 18 (1977): 435-44.

Squires, D., Grafton, R.Q., Alam, M.F. and Omar, I.H., 'Technical Efficiency in the Malaysian Gill Net Artisanal Fishery', *Environment and Development Economics*, 8 (2003): 481-504.

Chapter 4

Understanding and Measuring
Capacity in Fisheries

Excessive fishing capacity is largely responsible for the degradation of marine fisheries resources, for the dissipation of food production potential and for significant economic waste, especially manifest in the form of redundant fishing inputs.
Dominique Gréboval Managing Fishing Capacity: Selected Papers on Underlying Concepts and Issues. (1999, p. iv)

4.1 Introduction

The ability of fishing fleets to harvest fish well in excess of existing catches, either supported by existing resource conditions or long-run desired resource levels, is an issue of international concern. Beginning in 1995, the United Nations Food and Agriculture Organization (FAO) and various member nations embarked on an ambitious plan to address excess capacity in fisheries. Article 6.3 of the Code of Conduct for Responsible Fisheries, adopted by the FAO in 1995, recommended that 'States should prevent overfishing and excess fishing capacity, and should implement management measures to ensure that fishing effort is commensurate with the productive capacity of the fishery resource and their sustainable utilization'.

Despite a thorough investigation of capacity and the methods for estimating capacity output in fisheries, the concepts and methods still remain vague, ill defined, or simply not understood by some fisheries managers. In this chapter we define and explain what is capacity output and closely related ideas of excess capacity, capacity utilization and variable input utilization. To understand these concepts, a review of the material covered in chapters 1 and 3 would be helpful, especially the notion of technical efficiency. In addition to explaining capacity concepts, we also briefly review various approaches to measuring capacity output and utilization in fisheries and provide specific fisheries applications.

4.2 Defining Capacity and Related Concepts

Capacity or capacity output can be defined in a number of different ways. A technological way of measuring capacity uses the relationship between inputs and outputs to measure capacity. It is defined as the maximum output level that can be produced given the fixed inputs (where the amount used can only be changed in the long run), the existing technology, the availability of variable inputs (where the amount used can be changed in the short run), and customary and usual operating procedures. In fisheries, any measure of capacity must also consider the effect of the level of the fish stock or abundance, and not only the technology and inputs used. For instance, with a much larger fish stock, but with the same technology and inputs, total landings will almost certainly be higher than with a much smaller fish stock. Another consideration when measuring capacity is whether it is defined as a short or a long-run concept. If the fixed inputs cannot be changed then the measure of capacity is a short run concept conditional on, or constrained by, the fixed inputs (e.g., vessel tonnage or hold capacity in fisheries). If the fixed inputs can all be changed then, presumably, it is possible to produce a higher maximum output and the measure of capacity would be different.

A related, but different, concept is capacity utilization. Unlike capacity that is measured in terms of output, capacity utilization is often measured as a ratio of current or observed output relative to a potential or capacity output. Capacity utilization is frequently confused with the term capital utilization that is the ratio of the current or observed capital stock (such as vessel size) to the 'ideal' level of capital. Only under special circumstances such that there is only one (and only one) fixed input, and where output increases by the same proportion as the increase in all inputs, will capacity utilization be the same as capital utilization. In fisheries, managers also frequently refer to the overcapitalization that arises whenever capital utilization exceeds 1.0 such that too much capital is being used relative to the ideal level. In general, economists prefer measuring capacity and capacity utilization because they provide broader measures of performance than simply considering only one fixed input, as is the case for capital utilization and overcapitalization.

Technological Concept of Capacity

To better understand the technological concept of capacity, consider a fisher who generates a harvest or output (Y) using only one input (V), as illustrated by Figure 4.1. For the moment we will ignore the effects of the fish stock on capacity, the existence of multiple inputs, and whether some inputs are fixed in the short run.

Figure 4.1 shows that initially harvest increases at a greater rate than the change in the variable input, as represented by the line segment OA. This implies that if the input increases by 10 per cent, output increases by greater than 10 per cent. The harvest level then increases in the same proportion as the input as shown by the line segment AB, where if the input is increased by 10 per cent, output is increased by 10 per cent. Lastly, output increases by a proportion less than the increase in the input as illustrated by line segment BC, where if the variable factor is increased by 10 per cent, output is increased by less than 10 per cent. This type of technology is referred to as a variable returns to scale because the proportional change in output from changes in the input varies with the level of output.

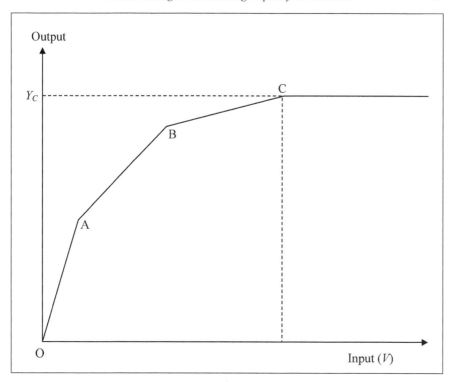

Figure 4.1 Technological Concept of Capacity

The line segments given by OABC represent a production frontier (see chapter 3) and graphs the most output or harvest that can be obtained from all possible levels of the input. If production took place beneath the frontier then the fishing vessel would be technically inefficient in the sense that it failed to produce the maximum output from a given set of inputs. Although all points on the frontier are technically efficient, only the point that corresponds to maximum possible output, given the input, is capacity output (YC). In other words, capacity output is defined as the maximum possible output that can be produced given the technology and the input level. Output cannot be expanded beyond YC, because of technological or production constraints. We stress that the measure of capacity output is purely a technological concept defined by the production frontier and, therefore, is independent of prices and costs. Thus whether it would be economically profitable to produce at capacity output cannot be resolved without further information.

Economic Concept of Capacity

An economic, as opposed to technological, measure of capacity and capacity utilization is more difficult to define because we have to compare current or observed levels of output to some 'ideal', rather than to some maximum level of output. The metric or unit of measurement used by economists for determining the ideal output can vary. It may be defined by costs of production, where the ideal or capacity output is that which minimizes costs (Morrison 1985), or it might be defined as the level of output that maximizes revenues (Färe et al. 2000a) or profits (Coelli et al. 2000).

The most commonly accepted economic concept of capacity output in the short run corresponds to the output level where short-run average cost and the long-run average cost curves are tangent to each other. This is illustrated in Figure 4.2 where the large U shape curve represents long-run average cost. This long-run curve maps the relationship between the average cost of production and output when all inputs (fixed and variable) can be changed. The long-run curve can be thought of as a 'lower envelope' of many smaller short-run average cost curves such that every point on the long-run average cost curve corresponds to a tangency with a short-run average cost curve. For example, points D, B and E represent tangent points for three different short-run cost curves and the long-run curve is constructed from a set of these and other tangent points. In other words, long-run average cost corresponds to the minimum average cost for every possible level of the fixed input.

The long-run average curve must always be equal to or lie below any short-run average curve because, in the long-run, all inputs can be varied which allows the firm to have at least as lower cost than if only a few inputs can be changed. The three smaller U-shape curves in Figure 4.2 are short-run average cost curves and represent the relationship between average costs and output when only the variable input can be changed. A short-run average cost curve is U-shaped to reflect the fact that total costs of production include a variable component (variable costs), which increases with output, and a fixed component (fixed costs) that does not change with output. Thus, initially when there is only a small amount of output, fixed costs tend to dominate total costs such that an extra unit of output will lower the average cost of production. As output increases, variable costs become more important because these costs also rise with output such that average cost will eventually get bigger as output increases.

The short-run average cost curve represents the cost-output relationship for a given level of a fixed input. The long-run average cost curve represents the cost-output relationship when all inputs (including the fixed input) can vary. For simplicity, we shall assume there is only one variable input, such as labor, and one fixed input, capital or vessel size. In reality there will be several inputs that skippers cannot vary in the short run that complicate the economic measurement of capacity. The points A, B and C in Figure 4.2 are all on the same short-run average cost curve. As a result, although they represent different levels of output and variable inputs, they all have the same capital stock or level of the fixed input. Short-run capacity output is given by Y_C^s because it corresponds to point B, or the point of tangency to the long-run curve. Thus short-run capacity output represents the minimum long-run average cost of production for the given level of the fixed capital stock. Short-run average cost could, in fact, be lower with a slightly larger level of output, but this would also require larger amounts of the variable input and, in the long run, this would generate a greater cost than if the capital stock were to be increased and the variable input left unchanged.

Long-run capacity output, defined by Y_C^*, is located at the minimum of the long-run average cost curve because it minimizes average costs when all inputs can be varied, including the capital stock. If the average cost of production is minimized when all inputs (fixed and variable) can be set at their optimal (cost minimizing) levels then there is no incentive to change the scale of production or the level of inputs. In other words, long-run capacity output is coincident with both minimum long-run average cost and the optimal scale of production.

At any point in time, however, we are in the short run and cannot adjust all our inputs to be at the optimal scale of production. Thus capacity output in the short run is constrained by the fixed level of input or capital stock. This has important implications in terms of capacity that can be visualized using the short-run average cost curve defined by the points A, B and C. In the short run, for output levels less than Y_C^s, such as output level YE which corresponds to average cost level A, there is excess capacity because it is possible to have a lower amount of the fixed input and be at a lower cost. This can be seen by comparing the average cost at point A with the short-run average cost curve that is tangent to the long-run average cost at the same level of output, YE. If the fisher were to produce output YE using the level of fixed input that coincides with the short-run average cost defined by the points A, B and C, she would be able to reduce her costs of production if she were to use smaller amounts of the fixed input. A lower amount of the fixed input would also put the fisher on a different short-run average cost curve defined by point D. The vertical distance between A and D represents the cost saving from using fewer units of the fixed input and an optimal amount of the variable input.

A similar analysis can also be undertaken when comparing output that exceeds short-run capacity at Y_C^s, such as output level YU which corresponds to average cost level C. The higher output level represents under capacity because it is possible to have a larger amount of the fixed input and be at a lower cost. This can be seen by comparing the average cost at point C with the short-run average cost curve that is tangent to the long-run average cost at the same level of output, YU. A larger amount of the fixed input would put the fisher on a different short-run average cost curve defined by point E. The vertical distance between C and E represents the cost saving from using more units of the fixed input and an appropriate amount of the variable input.

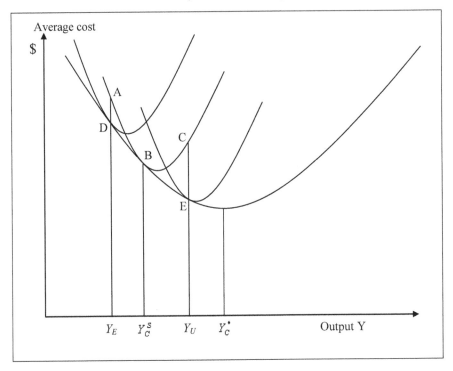

Figure 4.2 Economic Concept of Capacity

As illustrated in Figure 4.2, the economic concept of capacity utilization may be greater than, equal to, or less than 1.0 if we use actual output compared to capacity output. If the economic concept of capacity utilization is greater than 1.0, there is under capacity, and if less than 1.0, there is excess capacity. If capacity utilization equals 1.0, production equals the capacity output.

Technological-Economic Concept of Capacity

Unfortunately, lack of good cost data prevents analysts from measuring economic capacity in many fisheries. For this reason we will develop a technological-economic measure of capacity. Although the technological-economic notion cannot be used to predict changes in capacity in response to changes in input and output prices (Kirkley and Squires 1999), it does give some implications about the economic behavior of fishers because it is based on observed data that reflect the actual decisions of fishers.

The technological-economic measure of capacity was introduced by Johansen (1968) and later developed by Färe et al. (1989) and has been referred to as the weak Johansen concept of capacity (Coelli et al. 2000). It is defined as the maximum output that can be produced per unit of time with existing plant and equipment (fixed inputs that cannot be changed in the short run), provided the availability of the variable inputs is not restricted. In the fisheries context it is the maximum harvest that a vessel can produce given the fixed factors of production (e.g., engine horsepower, vessel size, vessel hold, and winch size), readily available variable factors (e.g., labor and fuel), resource conditions, the state of the technology, and customary and usual operating procedures.

Harvesting Capacity

Many fishery managers would define capacity output as either the maximum output level that can be produced from a given set of inputs, or the minimum level of inputs used to produce a given output level. We call this measure harvesting capacity to distinguish it from our technological, economic and technological-economic concepts of capacity defined previously. Although harvesting capacity, the minimum level of inputs required to produce a given level of outputs, is not equivalent to our previous measures of capacity, it is widely used because fishery managers often wish to know the minimum number of vessels necessary to harvest a given total allowable catch (TAC), provided that all vessels operated at full capacity (Kirkley et al. 2004a). However, given that fishers have a number of fixed inputs which cannot be simply aggregated in terms of the number of vessels, harvesting capacity can be a misleading measure of capacity output in a fishery.

Excess and Under Capacity

Excess capacity arises whenever a firm or industry has the capability to produce more than it is actually producing or could produce, if it was operating efficiently. It is defined as the difference between the output level a firm or industry could produce if production were efficient and inputs were fully utilized, and the output level actually produced for a given set of input and output prices at a point in time for a given technology. Thus if the maximum potential output exceeds the output level actually produced there is excess capacity. By contrast, if a firm or industry does not have the productive capability to produce a given output level, there is a shortage or under capacity.

Excess capacity is a common phenomenon in many industries at different points in time. In fisheries, excess capacity is often considered as undesirable because it implies too much idle capital, and thus, wasted resources, given existing catch levels (Kirkley et al. 2004a). Another way of thinking about excess capacity is that the current output is too low to fully utilize the existing capacity base. Although excess capacity is a concern in fisheries where increased output of one vessel reduces the catch to all other vessels, this is not the case in many other industries. For example, a farmer with too much equipment for size of the farm may liquidate or sell off assets in order to downsize operations, or increase the size of the farm, or use the equipment for other activities other than farming. In the case of fisheries, however, excess capacity may pose more serious problems (Pascoe et al. 2004) that represent a potential 'overhang' which could have adverse consequences for both economic returns and sustainability of fish stocks. This overhang may last a long time in fisheries because gear, and even vessels, are often highly specialized to particular species, types of fishing and locations which makes it difficult to shed excess capacity once it occurs.

Another issue with excess capacity is the 'lumpiness' of fixed inputs because a fishing vessel cannot be easily sold off in pieces, except for maybe the electronics, or used for purposes other than commercial fishing. Excess capacity in fisheries is also frequently a symptom of poor management whereby managers restrict some inputs (especially season length) to prevent biological over-fishing, but the underlying incentive to 'race to fish' is unchanged. As a result, fishers substitute to unregulated inputs to catch the limited fish available before someone else does (Kompas et al. 2004). Although this might be beneficial to an individual fisher if she succeeds in capturing a greater share of the catch, it results in too many and the wrong sort of inputs being used.

Capacity Utilization

Capacity utilization (CU) is the ratio of technically efficient output to capacity output, but is typically measured as the ratio of actual or observed output to capacity output. It has also become common in fisheries to compute the inverse of capacity utilization, i.e., 1/CU. The technically efficient output is the frontier output (see chapter 3) or the maximum potential output for a given level of inputs, both variable and fixed. By contrast, technological-economic capacity output is the maximum potential output, given fixed inputs, but allowing the variable factors to increase until output is limited only by the fixed inputs and technology.

When measured as the ratio of observed or actual output to capacity output, CU incorporates both technically inefficiency and inadequate utilization of the variable inputs. In other words, actual output may be less than capacity because a vessel is not operating at the frontier and is technologically inefficient, or it might be because it is constrained by the level of the fixed factors, or for both reasons. When CU is measured as the ratio of technically efficient output to capacity output, it provides a measure of the inadequate utilization of the variable inputs.

To illustrate the difference between the two possible measures of CU we can use a fisheries example. Suppose a fishing vessel spends 150 days of fishing per year to catch 100 tons of fish, and each fishing trip takes one day so that its catch per fishing day/trip is 0.67 tons (100.5 tons/150 days). We assume that the vessel is technically efficient and that hold size is the only fixed factor that limits the landed catch of the vessel and equals 0.70 tons per fishing trip. In other words, no matter how successful the skipper and crew are at catching fish, over the course of one fishing trip the most fish they can bring back to port is 0.70 tons. This hold size provides an overall maximum output or catch for the vessel that we define as capacity output. If, as assumed, the vessel is at sea for 150 days per year then the capacity output is 105.0

tons (150 trips × 0.70 tons per trip). Consequently, CU or the ratio of the technically efficient output to the capacity output is 0.96 (100.5 tons/105.0 tons).

To realize the capacity output, however, the vessel operators would have to catch an extra 4.5 tons (105.0 tons less 100.5 tons) of fish which requires an extra 6.72 days or trips (4.5 tons/0.67 tons per trip = 6.72 days/trips). In other words, because the vessel lands slightly less than the capacity output per trip (0.03 tons per trip), to produce the capacity output it would have to increase its variable input, in this case the number of fishing trips, to be able to reach annual capacity output. Thus an increase in the number of fishing days/trips from 150 to 156.72 allows the vessel to have a full utilization level of the variable input – the number of days fished or the number of fishing trips.

Consider the same example, but now let us assume the vessel is technically inefficient. In this case, an inefficient level of production must be less than 0.67 tons per day. For illustration purposes, assume that inefficient production equals 0.50 tons per day or fishing trip. The total catch per year therefore equals 75 tons (150 trips × 0.50 tons per trip) or 30 tons less than annual capacity output of 105 tons. CU or the ratio of observed (rather than efficient) output to capacity output is thus 0.71 (75 tons/105 tons). The question is whether or not the vessel is not producing the capacity output because of inefficiency, inadequate usage or variable inputs, or both. If it were not known that the vessel production were inefficient, we might conclude that the full-utilization level of days or effort would require an extra 60 trips or days (30 tons/0.5 tons per trip = 60 days/trips). In fact, failure to be on the production frontier requires an extra 23.28 days/trips, and failure to be at capacity output while on the frontier contributes a further 6.72 days/trips.

This example can also be illustrated graphically in Figure 4.3. If the vessel were technically inefficient it would produce beneath the production frontier and generate a catch level of Y^1 units of output using V^1 units of variable input. If the vessel were technically efficient and used the same level of the variable input (V^1) it could land a greater amount of fish, denoted Y^2. Capacity output, denoted by Y^3, requires both technical efficiency and a greater usage of the variable input (fishing trips). Expanding the use of the variable factor beyond V^2, however, does not increase output because the fixed input (hold size) limits production to Y^3.

If capacity utilization, CU, is measured by the ratio of observed output to capacity output, CU equals Y^1/Y^3. If CU is measured as the ratio of technically efficient output to capacity output, CU then equals Y^2/Y^3. The technically efficient measure of CU indicates a need to increase variable input usage from V^1 to V^2. The CU measure, based on observed output to capacity output, would imply that a producer should increase variable input usage from V^1 to V^2 to realize the capacity output, but without improvements in technical efficiency the capacity output cannot be achieved.

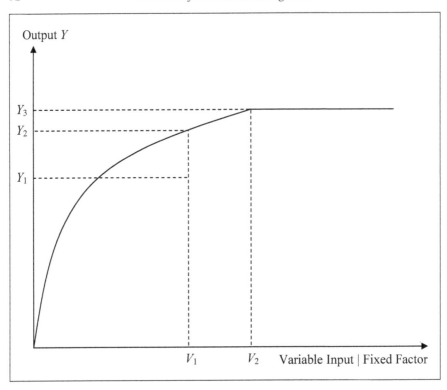

**Figure 4.3 Capacity Utilization in Terms of Observed and Technically
            Efficient Output**

Optimal Variable Input Utilization

A remaining concept important for understanding capacity is the optimum rate of
variable input utilization, or the full-utilization rate of variable inputs. This is a
measure of the level of variable inputs required to produce the capacity output. A
general measure of the variable input utilization, provided by Färe et al. (1994), is
the ratio of the optimum level of a variable input (i.e., the level required to produce
the capacity output) to the actual usage of a variable input used. If this ratio exceeds
1.0 in value, it implies that producers are using too little of a variable input, and if
it is less than 1.0 in value, it implies that producers are using too much of a variable
input. A value of 1.0 indicates that a producer is fully utilizing the variable input.
In Figure 4.3, the full variable input utilization rate equals the ratio of V^2 to V^1,
which provides a measure of the proportion by which variable input usage should be
increased to produce the capacity output.

Aggregation and Fleet-wide Measures of Capacity

We have examined capacity from the perspective of a single vessel or firm. To provide a fleet-wide measure of capacity, we must aggregate from individual vessels to all vessels in the fleet. Typically, aggregation places a lower bound on fleet-wide capacity because it is greater than or equal to the capacity output summed over all vessels in a fleet. This is because the existing use of inputs across the fleet is not necessarily optimal or the best allocation in terms of maximizing the total harvest. In other words, capacity output per operator/vessel summed over all operators/vessels is less than or equal to capacity output for a fleet.[1]

4.3 Capacity Output and Capacity Utilization

Much of our discussion on capacity is applicable to most industries, but in the fisheries context we must also explicitly account for 'environmental' variables such as the size of the fish stock. What can be harvested using a given set of inputs in a fishery clearly depends on how many fish there are in the sea. Thus, no meaningful measure of capacity in fisheries is possible without some reference to the size of the fish stock, and possibly other important environmental variables beyond the control of any individual fisher. In the fisheries context, therefore, comparisons of potential or capacity output depend on the underlying biological dynamics.

We can incorporate fish stock into measures of capacity using a production function which transforms or 'maps' inputs used by fishers, and also environmental variables, into an output. Typically, output is defined as Y and thus the production function is defined as

$$Y = f(V, K; S) \tag{4.1}$$

where V represents the variable inputs that can be changed in the short run, K represents fixed inputs (such as the vessel size) that can only be changed in the long run, and S represents environmental variables (such as fish stock size) that cannot be determined by an individual fisher. Fishing effort is some combination of the V and K inputs while S affects productivity, or effective catch.

The production function represents the maximum amount of output it is technologically possible to produce for a given level of inputs, with external factors controlled for by environmental variables denoted by S. A fisher that produces at the frontier defined by the production function is, by definition, technically efficient. Production beneath the frontier represents technical inefficiency. Technical efficiency ensures maximum output for a given level of inputs, but does not ensure a vessel is at capacity output, which requires the optimum use of inputs.

---

1   Färe et al. (2000b) provide a discussion on estimating capacity for a fishing fleet and also give an illustration of how to estimate fleet-wide capacity.

Figure 4.4 provides a way of illustrating the importance of optimal use of inputs, technical efficiency and environmental variables on capacity output. For expositional purposes, we assume that the fixed inputs can be aggregated into one capital stock and that the level of this capital stock is fixed in the short run at $\bar{K}$. The individual fisher can control the level of the variable input (V) up to a maximum amount determined by the regulator of the fishery, but cannot change the size of the fish stock, denoted by S. The upper two curves in Figure 4.4 denoted by Y_{TE0} and Y_{TE1} represent two production frontiers of technically efficient output that define the maximum possible output given the use of variable input, the fixed capital stock and two different sizes of the fish stock (S_0 and S_1). The lower curve, denoted by Y_A, represents an 'average' relationship between the variable input and output given the fixed capital stock and fish stock, but it does not represent best practice and is thus beneath the technically efficient frontier.

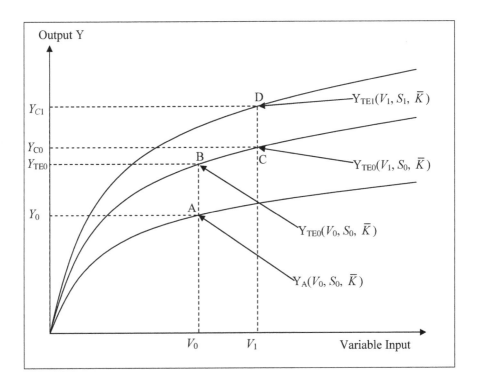

Figure 4.4 Capacity and the Production Frontier

In Figure 4.4, point A is technically inefficient because with the same amount of the variable input (V_0), fixed input ($\bar{K}$) and fish stock (S_0) it is possible, using the best harvesting practices, to produce a greater level of output at point B. The ratio B/A is called the efficiency score for observation A where the amount the score exceeds one indicates how much output could have increased if all measured inputs had been used in a technically efficient way.

Capacity output represents an ideal, but nevertheless obtainable level of output if the fisher were technically efficient and operating at the appropriate scale of production. As is common in many fisheries, inputs are often controlled by regulators to prevent the total allowable catch from being exceeded. In Figure 4.4, we assume that the maximum possible or available use of the variable input, given existing fisheries regulations, is V_1. Thus capacity output in the short run (capital stock is fixed at $\overline{K}$) is given by point C on the technically efficient production frontier for fish stock size S_0. If the fisher were able to use the variable input in a greater quantity than V_1 then the capacity output would be even higher.

A measure of capacity utilization, if we knew the technically efficient production frontier, would be the ratio B/C and if we used the average or observed production point A, would be defined by the ratio A/C. The ratio B/C is the preferred measure because it represents only the difference in capacity utilization due to use of the variable input and is purged of the effect of technical inefficiency. Sometimes the capacity utilization measure is given as the proportional increase in output required to reach capacity output which is the inverse of capacity utilization, i.e., C/B = 1/CU.

In a fishery the output that can be produced, or harvest, is also determined by the size of the fish stock. Consequently a measure of capacity output is not independent of stock size, or environmental and other variables that might influence the total catch. This is shown in Figure 4.4 by point D which has a higher level of output than point C despite the identical use of inputs (V_1 and $\overline{K}$) because it has a higher fisher stock (S_1 instead of S_0). By explicitly considering environmental variables such as fish stock size we can also measure what output might be possible with different stocks sizes or a target stock level. Imputing such a potential output level for a different stock size requires some estimate of the marginal product of S, or the extra output associated with a marginal change in the stock.

We have limited our discussion to a single output, but multiple outputs may also be accommodated in measures of capacity output and utilization. This is often done through the estimation of a distance function, rather than a production function, which refers only to a single-output technology. A difficulty when calculating capacity output in a multi-output setting is what assumption to make regarding the composition of outputs as one output changes.

If fishers cost minimize or profit maximize, and if suitable economic data is available, we can also generate an economic measure of capacity output. As with the technological-economic measures presented in Figure 4.4, it also requires an understanding of short-run constraints facing fishers. For example, if a fisher is unable to change all her inputs (such as vessel size) then the costs are likely to be higher than if all inputs can be optimized in the production process. Moreover, fishers may face regulatory constraints that prevent them from reaching the cost-minimizing bundle of inputs. Despite the differences in approach, the technological-economic and economic approaches are not necessarily inconsistent.

To better understand the relationship between technological and economic measures of capacity we can consult Figure 4.2. The short-run average cost curve that includes points A, B and C all have the same capital stock ($\bar{K}$) that is fixed in the short run. A move from point A, which represents a lower level of output, to point B indicates that output has increased and implies the use of the variable input must have increased if the capital stock and fish stock remain unchanged. These two production points (A and B) can be represented in a technological-economic notion of capacity by two different vessels producing different outputs, but which are the same size ($\bar{K}$). Short-run capacity output in a technological-economic sense would be given by the vessel at point B, that also coincides with the ideal or economic capacity output which minimizes long-run average cost for the given capital stock. Capacity utilization for the vessel at point A would therefore be the ratio of the output coincident with this level of production (Y_E) over capacity output (Y_C^s). Thus even without economic data on costs it is still possible using observed inputs and output data from different vessels to construct a frontier and thereby impute the maximum output produced for a given capital level – capacity output.

A potential problem, however, arises if the variable input is unconstrained and it is possible to produce an even greater output than Y_C^s, as represented by point C and output Y_U. In this case, a vessel producing at point C in Figure 4.2 might appear to be the best practice or most productive vessel because the ratio of output to the fixed input ($Y/\bar{K}$) is the highest (relative to points B and A) despite the fact that C is not an economically optimal point of production. In this case, if a vessel were producing at point C it would have too small an amount of the capital stock to be economically efficient because its short-run cost curve is greater than the long-run average cost curve at this level of output.

4.4 Methods for Measuring Capacity

There are several methods for estimating and assessing capacity output and capacity utilization in commercial fisheries, which are reviewed by Pascoe et al. (2003). These include the peak-to-peak method (Klein 1960), which compares actual output to highest observable output taking into account differences in technology, fixed inputs and environmental variables, the stochastic production frontier (described in chapter 3), and data envelopment analysis (DEA). A comparison of these methods is provided by Kirkley et al. (2004b). In this chapter we restrict our discussion to the most widely used method to measuring capacity in commercial fisheries, data envelopment analysis.

Data Envelopment Analysis

DEA is a mathematical programming method that is used to obtain a measure of efficient and capacity output that can be used for comparison to observed output levels. Charnes et al. (1978) originally introduced DEA to the operations research literature and it has subsequently been applied in many different contexts, including in fisheries and the measurement of capacity.

The method essentially 'envelops' the data or the observed input-output combinations from a fishing fleet to create a multiple-dimensional frontier. The advantage of DEA analysis is that it easily accommodates multiple outputs and inputs, while also allowing the imposition of constraints on production processes. As originally developed, DEA is non-parametric and non-stochastic in that deviations from a frontier are supposed to occur due to individual inefficiencies, or constraints and not due to random events. This is viewed as its main weakness although some methods have recently been developed to account for randomness or 'noise' in the data.

Färe et al. (1989) used DEA to estimate capacity output, capacity utilization, and variable input utilization. Färe et al. (1989) initially proposed the DEA framework for assessing capacity when data were limited to only input and output information – the most commonly available data in developed commercial fisheries. Although such estimates of capacity output do not directly use economic data, they are based on actual data that reflects economic optimizing behavior by fishers. More recently, Färe et al. (2000a) and Färe and Grosskopf (2004), have provided DEA frameworks for estimating capacity that correspond to revenue or profit maximization and cost minimization.

The original DEA approach to estimating capacity output calculates maximum potential output given the fixed factors and environmental variables, but with no constraints on variable inputs. It is equivalent to the previously defined notion of technological-economic or weak Johansen notion of capacity. The method involves solving an optimization problem that maximizes output subject to constraints that include the actual levels of the fixed inputs, all variable inputs are fully utilized, levels of environmental variables (such as stock size) and general characteristics we would expect of a harvesting technology. This is done by using input-output data from a sample or population of vessels to construct a frontier that represents maximum output given the input and environmental constraints faced by fishers. The optimization procedure calculates capacity output for each observation by multiplying it by a weight, usually defined as θ, that is a measure of technical efficiency. If $\theta = 1.0$ then observed output equals capacity output, and if $\theta > 1.0$ then capacity output equals θ multiplied by the observed output level.

In addition to obtaining an estimate of capacity, the DEA approach using input and output data can be used to estimate capacity utilization. It can be calculated as the ratio of observed output to capacity output, but such a measure would include both technical inefficiency and less than full utilization of variable inputs which makes it difficult to determine to what extent inefficiency, rather than inappropriate utilization of inputs, is causing less than capacity output. An unbiased and alternative measure of capacity utilization, due to Färe et al (1989), calculates an efficient level of output. In this way a measure of capacity utilization is obtained without bias due to technical inefficiency. This unbiased method for calculating capacity utilization is bounded between 0 and 1.0. A value less than 1.0 indicates that production is at less than the full input utilization level required to produce the capacity output. In other words, output could be expanded with no additional inputs. A value of 1.0 indicates that if a fleet or vessel were operating efficiently and using the levels of fixed and variable factors required for producing the capacity level, it would be possible to produce the capacity output level.

A requirement for capacity output is that there is full utilization of the variable inputs and the vessel is technically efficient. The full-utilization level of the variable inputs can also be calculated from the solution to the input-output DEA problem as the ratio of the level of the ith variable input needed to produce capacity output to observed level of the ith variable input. A value greater than 1.0 indicates variable input usage must be increased to achieve capacity output while a ratio less than 1.0 implies the use of the variable input should be reduced.

4.5 Fisheries Applications

Using a data set that contains trip-level observations on fishing activities by nine US northwest Atlantic sea scallop vessels in 1990, we can illustrate the estimation and assessment of capacity. In the data, catch is measured in terms of pounds of meat weights, which is the landed product form for sea scallops; the fixed factors are vessel tonnage, engine horsepower, and dredge size; stock abundance; and the variable factors are days at sea and number of crew per trip. Stock abundance is the geometric mean of all samples, for a given resource area, of the number of baskets per standard tow for the last tow by vessels. Table 4.1 provides summary statistics of the nine vessels where a large coefficient of variation (standard deviation/mean) indicates a large spread or variation in the listed variable.

Table 4.1 Summary Statistics for Nine US Northwest Atlantic Sea Scallop Vessels

Variable	Minimum	Maximum	Mean	Coefficient of Variation
Gross Registered Tonnage	124.0	190.0	141.5	0.16
Horsepower	520.0	620.0	540.5	0.08
Days Per Trip	2.0	26.0	17.6	0.28
Crew Size Per Trip	6.0	13.0	8.9	1.85
Catch Per Trip	146.0	20,508.0	9913.0	0.45

Using DEA we can solve for capacity output, technically efficiency output, excess capacity (capacity output less observed output) and capacity utilization (ratio of technically efficient output to capacity output). The mean values per fishing trip for each of these measures is provided in Table 4.2 by vessel. The results show excess capacity for all vessels and the estimated total excess capacity for the fleet as a whole equals about 1.0 million lbs of meats. In other words, the fleet has the potential to produce 79 per cent more meat per trip (2.3 million lbs) than the amount actually harvested (1.3 million lbs).

To harvest the trip capacity output of 2.3 million lbs the fleet would need to improve technical efficiency, and also increase variable inputs as measured by both days at sea per trip and crew size. If technical efficiency were improved, but variable inputs left unchanged, fleet-wide production would also increase by about a third rather than the 79 per cent required to reach capacity output.

DEA can also be used in a multi-species context to provide individual species measures of capacity and capacity utilization. Using data from 1998, 1990 and 1991 from the Nova Scotia mobile gear fleet that fishes off the east coast of Canada, Dupont et al. (2001) were able to examine the effects, by species, of introducing individual transferable quotas (ITQs) into the fishery in 1991. A summary of their findings for the three species (cod, haddock and pollock) included in the ITQ program is provided in Table 4.3.

Table 4.2 Estimated Mean Value Per Trip of Capacity and Related Concepts

Vessel	Gross Registered Tonnage	Engine Horsepower	Observed Catch (100s of pounds)	Capacity Catch (100s of pounds)	Excess Capacity (100s of pounds)	Efficient Catch	Capacity Utilization	Observed Variable Input Usage		Optimum Variable Input Usage	
								Days	Crew Size	Days	Crew Size
1	181.0	620.0	12,563.3	19,350.0	6,786.7	16,692.6	0.87	20.2	9.2	19.3	11.2
2	125.0	520.0	7,356.7	15,493.2	8,136.5	10,186.9	0.65	18.9	8.7	19.6	9.0
3	190.0	620.0	12,249.6	18,987.4	6,737.8	16,581.0	0.88	20.4	9.4	19.4	11.0
4	124.0	520.0	9,242.1	15,813.5	6,571.4	12,127.6	0.77	19.1	9.1	20.7	9.0
5	130.0	520.0	9,824.4	18,121.1	8,296.7	13,964.7	0.78	17.6	9.6	19.4	10.2
6	135.0	520.0	11,469.7	18,346.4	6,876.7	13,929.9	0.77	18.5	8.7	18.9	10.9
7	129.0	520.0	9,132.8	16,853.5	7,720.7	11,567.3	0.69	16.2	8.1	18.9	9.8
8	137.0	520.0	8,668.8	18,684.6	10,015.8	12,261.5	0.65	14.9	9.1	19.1	10.9
9	131.0	520.0	9,549.8	18,298.7	8,748.9	12,104.5	0.66	16.7	8.4	19.5	10.2
Fleet[a]	141.5	540.5	1,308,520	2,341,735	1,033,215	1,735,806	0.74	2,323	8.9	2,563	10.2

[a] Aggregate fleet-wide values based on summation of observed catch and estimates over all observations. Fleet wide crew size is in terms of mean values per trip.

Table 4.3 Capacity and Excess Capacity in the Nova Scotia Mobile Gear Fishery

Year/Species	Capacity (metric tonnes)	Excess Capacity (metric tonnes)
Cod capacity		
1988	83,522	40,725
1990	89,653	50,331
1991	61,682	44,554
Haddock capacity		
1988	40,613	24,448
1990	25,262	14,053
1991	18,904	18,904
Pollock capacity		
1988	30,788	16,534
1990	24,853	10,596
1991	20,295	10,456

Source: Dupont et al. (2001, Table 6).

The results indicate substantial declines in excess capacity from 1988 to 1991 for haddock and pollock, but no change in cod excess capacity. These declines were caused by a fall in the capacity per day per vessel, and also a reduction in the total number of vessels fishing, that is attributed to the introduction of ITQs.

4.6 Conclusions

Capacity output is defined from both a technological-economic and a more rigorous economic perspective. The technological-economic concept is the maximum potential output given the technology, fixed factors of production, no restrictions on variable input usage, and customary and usual operating procedures. The economic concept is defined as an ideal output that corresponds to a tangency between the short- and long-run average cost curves. Most fisheries of the world, however, lack the cost data necessary to estimate the economic concept of capacity output. A related concept is capacity utilization which is typically defined as the ratio of observed output to capacity output. However, a better measure is to define capacity utilization as the ratio of the technically efficient output to capacity output. This allows fishery managers to determine to what extent inadequate usage of variable inputs, rather than technical inefficiency, is contributing to the failure to produce at capacity output.

When applying capacity measures in fisheries it is important to explicitly consider the effects of environmental variables, such as stock size, on output. Data envelopment analysis (DEA) is a mathematical programming approach that can account for environmental variables when measuring capacity and is also the most widely used method in fisheries. This approach is not without its weaknesses, but does provide a readily available method for estimating capacity output, technically efficient output, capacity utilization and excess capacity.

The information from capacity studies provides valuable signals to managers about the effect of regulations and other factors on fisher performance. For instance, it provides an indication of how successful managers have been at preventing 'effort creep' and the build up of a capacity 'overhang' that might threaten the sustainability of the fishery. Such information should feed back and result in better decisions to improve economic performance, and help ensure the sustainability of fish stocks.

References

Charnes, W., Cooper, W., and Rhodes, E., 'Measuring the Efficiency of Decision Making Units', *European Journal of Operational Research*, 2/6 (1978): 429-44.

Coelli, T., Grifell-Tatje, E., and Perelman, S., Capacity Utilization and Short-run Profit Efficiency (Manuscript, Presented at the North American Productivity Workshop, June 2000).

Dupont, D.P., Grafton, R.Q., Kirkley, J., and Squires, D., 'Capacity Utilization Measures and Excess Capacity in Multi-Product Privatized Fisheries', *Resource and Energy Economics*, 24 (2001): 193-210.

Färe, R. and Grosskopf, S., *New Directions: Efficiency and Productivity* (Boston: Kluwer Academic Publishers, 2004).

Färe, R., Grosskopf, S., and Kirkley, J., 'Multi-Output Capacity Measures and Their Relevance for Productivity', *Bulletin of Economic Research*, 52/2 (2000a): 101-12.

Färe, R., Grosskopf, S., Kirkley, J., and Squires, D., Data Envelopment Analysis (DEA): A Framework for Assessing Capacity in Fisheries When Data are Limited (Presented at the IIFET X conference, July 2000b).

Färe, R., Grosskopf, S., and Kokkelenberg, E., 'Measuring Plant Capacity, Utilization and Technical Change: A Nonparametric Approach', *International Economic Review*, 30/3 (1989): 656-66.

Färe, R., Grosskopf, S., and Knox Lovell, C.A., *Production Frontiers* (New York: Cambridge University Press, 1994).

FAO (Food and Agriculture Organization), *Code of Conduct for Responsible Fisheries* (FAO, Rome, 1995).

Gréboval, D., *Managing Fishing Capacity: Selected papers on Underlying Concepts and Issues* (FAO Fisheries Technical Paper, no. 386, Rome, 1999).

Johansen, L., 'Production Functions and the Concept of Capacity', Recherches récentes sur la fonction de production (Namur, Centre d'Etudes et de la Recherche Universitaire de Namur, 1968).

Kirkley, J. and Squires, D., 'Measuring Capacity and Capacity Utilization in Fisheries', in D. Gréboval (ed.), *Managing Fishing Capacity: Selected Papers on Underlying Concepts and Issues* (FAO Fisheries Technical Paper, no. 386, Rome, 1999).

Kirkley, J., Morrison Paul, C.J., and Squires, D., 'Deterministic and Stochastic Capacity Estimation for Fishery Capacity Reduction', *Marine Resource Economics*, 19 (2004a): 271-94.

Kirkley, J., Walden, J., and Waters, J., 'Buyback Programs: Goals, Objectives, and Industry Restructuring in Fisheries', *Journal of Agricultural and Applied Economics*, 36/2 (2004b): 333-45.

Klein, L.R., 'Some Theoretical Issues in the Measurement of Capacity', *Econometrica*, 28 (1960): 272-86.

Kompas, T., Che, T.N., and Grafton, R.Q., 'Technical Efficiency Effects of Input Controls: Evidence from Australia's Banana Prawn Fishery', *Applied Economics*, 36 (2004): 1631-41.

Morrison, C.J., 'Primal and Dual Capacity Utilization: An Application to Productivity Measurement in the U.S. Automobile Industry', *Journal of Business and Economic Statistics*, 3 (1985): 312-24.

Pascoe, S., Gréboval, D., and Kirkley, J., 'A Framework for Capacity Appraisal in Fisheries', in D. Gréboval (ed.), *Measuring and Appraising Capacity in Fisheries: Framework, Analytical Tools, and Data Aggregation* (FAO Fisheries Circular, no. 994, Rome, 2004).

Pascoe, S., Kirkley, J., Gréboval, D., and Morrison-Paul, C.J., *Measuring and Assessing Capacity in Fisheries: 2 Issues and Methods* (FAO Fisheries Technical Paper, 433/2, Rome, 2003).

Chapter 5

Measuring Productivity and Decomposing Profits in Fisheries

There is a lad here, which hath five barley loaves and two small fishes: but what are they among so many?

The Bible (Authorised version, 1611), John 5:12.

5.1 Introduction

Productivity can be measured in many different ways, but is typically represented as the ratio of some output or collection of outputs to some input or collection of inputs. Its importance in fisheries is that it provides an indicator of economic performance or changes in fish stocks and abundance, or both.

To better understand the relationship between an output and an input, economists define what is called a *production function*. It can be used to plot how much output can be produced at different levels of input. If this relationship were fixed, or linear, the production function would be represented by a straight line if output were plotted on the vertical axis and if input were plotted on the horizontal axis. In reality, how much output can be produced per unit of input will change depending on the amount of input used. For example, it may be possible that increasing the amount of the input (such as labor) can lead to an even greater proportional increase in output as the workers specialize in their individual tasks and climb the 'learning curve'. Eventually, without increased amounts of other inputs, employing more and more workers will be subject to *diminishing returns* such that further increases in the single input will lead to proportionately smaller and smaller increases in output. It may even be possible for output to decline with successive increases in the number of people employed if they were to impede each other in their work.

Figure 5.1 illustrates the case of how at an intermediate use of the input (points between D and C) output increases by a greater proportion than the input, but at some point (point C) diminishing returns occur. We define productivity as the ratio of the output divided by the input used. Thus the higher the ratio, the more productive is the firm, all else equal. The productivity of any point on the production frontier is therefore the slope of a line (the vertical distance over the horizontal distance) drawn from the origin to the production frontier at that particular input-output combination. Point C represents the input-output level with the highest productivity because the vertical distance (measured by output, or the distance E to C) divided by the horizontal distance (measured by input, or the distance O to E) is greatest at this point. If diminishing returns were to start from the very beginning (which is *not* the case in Figure 5.1) and were to continuously fall with each successive unit of input, the slope of the production frontier would also decline continuously and, thus, the highest productivity point would occur with the very first unit of the input.

In Figure 5.1 all points on the production frontier represent *technically efficient* levels of production such that for any amount of input, the maximum output is produced (see chapter 3 for a fuller explanation). Firms that produce an input-output combination beneath the frontier, such as at point A, are technically inefficient because for the same input they could (if they were technically efficient) produce a higher output such as point B. The larger is the gap between A and B the greater the technical inefficiency of the firm.

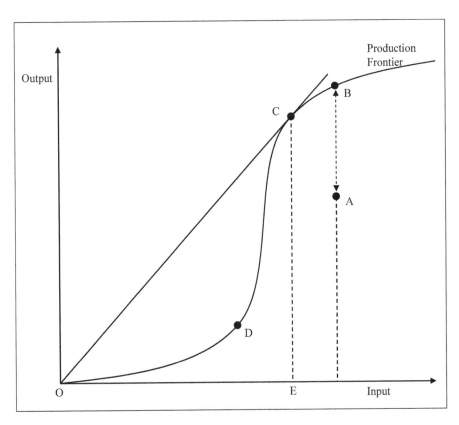

Figure 5.1 Productivity, Efficiency and the Optimal Scale of Production

Figure 5.1 also shows that productivity is increasing in the amount of the input up until point C, beyond this point productivity declines. In other words, there is an optimal *scale of production* that maximizes productivity defined by point C. Too much input (beyond point C) or too little input (from the origin up to, but not including point C) means that the productivity of the firm will not be at the maximum possible given the available technology. A key point to understand is that simply being technically efficient (producing on the production frontier) does *not* maximize overall productivity, but instead maximizes productivity only for a given input-output combination. For example, point D is technically efficient as it is on the production frontier, but it has a lower productivity than point C.

So far we have assumed that the production frontier is fixed, but in reality the relationship between how much can be produced from a given level of input varies over time. The production frontier could vary because of technical change that allows fishers to catch more fish for a given level of effort, or it could change because of fluctuations in the size of the fish stock, or for environmental reasons. Figure 5.2 illustrates how the production frontier could shift upwards from period t to period t + 1 from either improvements in technology, an increase in the fish stock, or both. In this illustration the maximum productivity level of productivity, denoted by point C, also shifts but is higher in period t + 1.

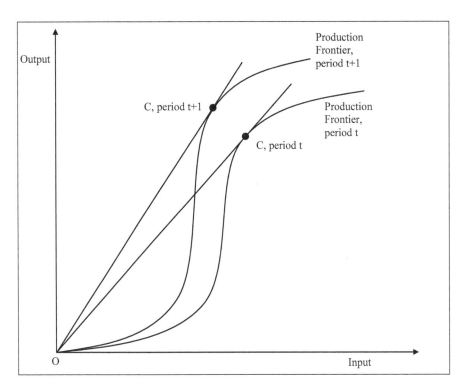

Figure 5.2 Change in Production Frontier and Productivity over Time

5.2 Productivity Measures in Fisheries

In fisheries we should be concerned with several measures of productivity. For example, we may want to track the performance of the total harvest (H) per number of fishing hours or some other measure of fishing effort (E) to obtain a measure of catch per unit of effort ($CPUE$). In Equation (5.1) the ratio of the total harvest in a fishery at time t to a measure of the fishing effort at time t and multiplied by a catchability coefficient represents a productivity index for the fishing fleet as whole at a particular point in time, i.e.,

$$CPUE_t = q\left(\frac{H_t}{E_t}\right) \tag{5.1}$$

where for ease of exposition we normalize the catchability coefficient such that $q = 1$.

A productivity index can also be constructed to compare different vessels at a given point in time, as shown in equation (5.2) for a vessel i,

$$\text{Productivity}_i = \frac{H_i}{E_i} \tag{5.2}$$

Productivity measures can also be derived over time and between vessels at a given point in time. Thus a comparison of the productivity between a vessel denoted by i, and another vessel denoted by j, can be defined by equation (5.3).

$$\text{Productivity Index (between } i \text{ and } j) = \frac{\dfrac{H_i}{E_i}}{\dfrac{H_j}{E_j}} = \frac{\dfrac{H_i}{H_j}}{\dfrac{E_i}{E_j}} \tag{5.3}$$

The productivity index in equation (5.3) is defined as the ratio of the two individual measures of productivity. If the ratio is greater (less) than one it implies that vessel i is more (less) productive than vessel j at that particular point in time. It can also, and equivalently, be defined as the ratio of an output index and an input index between vessels i and j, as shown in the far right-hand side of (5.3).

A productivity index can also be used to measure changes in productivity over time. For example, an index that measures changes in $CPUE$ over time (between periods t and $t+1$) for a fishery is given by equation (5.4).

$$\frac{CPUE_{t+1}}{CPUE_t} = \frac{H_{t+1}}{H_t} \Big/ \frac{E_{t+1}}{E_t} \tag{5.4}$$

From equation (5.4) it can readily be deduced that if the growth in the harvest *equals* the growth in fishing effort between the two periods then the growth in productivity is zero, and the productivity index will equal 1.0. In other words, for productivity to change over time it must be the case that proportional change in the output measure is *different* to the proportional change in the input measure.

Productivity measures are useful indicators of both economic and biological performance. For instance, declines over time in overall productivity, as measured by changes in *CPUE*, may be an indicator of declining fish stocks or abundance as it implies that increasing amounts of effort are required just to catch the same amount of harvest as in the past. Rather than use a productivity measure to provide information about stocks, regulators interested in a better understanding of fisher performance should also use indicators of stock size or abundance (*B*) that are derived *independently* of the measure of productivity (Squires 1992; 1994). In this way, we can calculate a stock-adjusted measure of *CPUE* over time, denoted by *SCPUE*, as in equation (5.5).

$$\frac{SCPUE_{t+1}}{SCPUE_t} = \left(\frac{H_{t+1}}{H_t} \bigg/ \frac{E_{t+1}}{E_t} \right) \times \frac{B_t}{B_{t+1}} \qquad (5.5)$$

By *not* accounting for the change in stock, the fishery manager may find that the productivity performance of the fishing fleet improves over the period if the harvest increases by a proportion greater than the proportional increase in fishing effort. This might erroneously be interpreted as improvements in fishing practices or efficiency, but it could also be due to a larger fish stock which allows harvesters to catch fish with much less effort. Adjusting for changes in the stock or abundance over the period allows the manager to discern what increases in productivity are due to fluctuations in stocks, and what may be caused by changes in overall fishing performance.

5.3 Output, Input and Productivity Indexes

When calculating productivity in (5.4) and (5.5) we assumed that we could add all the output or catch across all vessels and treat fishing effort as simply one input. In reality, there may be many species of fish caught with a variety of different inputs. Aggregating such outputs and inputs for individual vessels, across vessels and over time requires that we assign 'weights' to the various output and input quantities. The weights we normally use are prices of the outputs and the inputs so that everything is in the same units whether it be dollars, Euros or pounds. This aggregation process also ensures that an output (input) that represents a large fraction of the total revenue (variable costs) is given a greater 'weight' than an output (input) of little monetary importance.

If we use output (input) prices as weights in an output (input) quantity index then how prices change over time will affect our measures of the quantity changes in outputs (inputs). Different indexes assign different weights when calculating an output or input quantity index. Two commonly used quantity indexes are the *Laspeyres* and the *Paasche index*, and a third index, the *Fisher index* which is a geometric mean of these two indexes. The Laspeyres quantity index with a fixed base or set of weights uses base-period prices to aggregate quantities while the Paasche quantity index with a fixed base or weights uses current-period prices.

To show how to calculate the three indexes, and thus aggregate across different outputs and inputs, let us assume a vessel catches three different species of fish denoted by $i = x$, y and z, and receives the following output prices p_x, p_y and p_z. The approach, however, is perfectly general to aggregate outputs and inputs over several vessels and over time. The respective prices and quantities harvested by the vessel in periods t, $t+1$ and $t+2$ are given in Table 5.1.

Table 5.1 Hypothetical Output Prices and Quantities in Periods t, $t+1$ and $t+2$

	Period t	Period $t+1$	Period $t+2$
Output Prices			
p_x	$10	$8	$7
p_y	$5	$9	$6
p_z	$6	$7	$5
Output Quantities			
q_x	1	2	3
q_y	2	1	2
q_z	3	2	4

The Laspeyres output index with a fixed base between periods t and $t+2$ is calculated as follows:

$$\text{Laspeyres Qty. Index}_{t,t+2} = \frac{\sum_{i=x}^{z} p_{it} \times q_{it+2}}{\sum_{i=x}^{z} p_{it} \times q_{it}} = \frac{\$10 \times 3 + \$5 \times 2 + \$6 \times 4}{\$10 \times 1 + \$5 \times 2 + \$6 \times 3} = \frac{64}{38} = 1.6842$$

$$(5.6)$$

where we use base period (period t) prices to weight the outputs in the two different periods of interest.

The Paasche index output index with a fixed base between periods t and $t+2$ is calculated as follows:

$$\text{Paasche Qty. Index}_{t,t+2} = \frac{\sum_{i=x}^{z} p_{it+2} \times q_{it+2}}{\sum_{i=x}^{z} p_{it+2} \times q_{it}} = \frac{\$7 \times 3 + \$6 \times 2 + \$5 \times 4}{\$7 \times 1 + \$6 \times 2 + \$5 \times 3} = \frac{53}{34} = 1.5588$$

(5.7)

where we use the latest or most current period (period $t+2$) prices to weight the outputs in the two periods of interest.

The Fisher index (named after Irving Fisher (1922)) is the geometric mean of the Laspeyres and Paasche index and is calculated as follows,

$$\begin{aligned}
\text{Fisher Qty. Index}_{t,t+2} &= \sqrt{\text{Laspeyres Qty. Index}_{t,t+2} \times \text{Paasche Qty. Index}_{t,t+2}} \\
&= \sqrt{1.6842 \times 1.5588} \\
&= 1.6203
\end{aligned}$$

(5.8)

The Fisher index is generally preferred because it has a number of desirable properties. For instance, the change in the Fisher index between an earlier and a later period is the reciprocal of the index from the later to the earlier period, while the product of a Fisher quantity index and a Fisher price index (an index that measures changes in prices while using quantities as weights) exactly equals a Fisher index that measures changes in nominal values (prices X quantities) over time. It also has the virtue of being able to accommodate zeros in the data.

Provided we have input quantity data and prices for the same vessel, and for the same time periods, we could also calculate an input quantity index in exactly the same way as we have done in equations (5.6) to (5.8) for output. In other words, we would use base-period input prices for weights in terms of the input quantities to calculate a Laspeyres input quantity index, and use current-period input prices to calculate a Paasche input quantity index. As in equation (5.8), the Fisher input quantity index is the geometric mean of the Laspeyres and Paasche quantity indexes.

If we have both an output quantity and an input quantity index, we can also calculate a productivity index between any two periods (say periods t and $t+2$). Such an index represents the ratio of the change in weighted outputs between the two periods over the change in the weighted inputs between the two periods, i.e.,

$$\text{Productivity Index}_{t,t+2} = \frac{\text{Output Qty. Index}_{t,t+2}}{\text{Input Qty. Index}_{t,t+2}}$$

If the change in output index exactly equals the change in the input index then the productivity index will equal 1.0, indicating no change in productivity between the two periods. A productivity index can be calculated as the ratio of Laspeyres, Paasche or Fisher output and input quantity indexes or as the ratio of some other type index, such as the Törnqvist index. For the purposes of directly measuring productivity in fisheries, a Fisher index is preferred over a Laspeyres or Paasche index because of its desirable properties. Thus a Fisher productivity index is the ratio of a Fisher output index over a Fisher input index. As already emphasized, for a productivity index to be as useful as possible we should also stock-adjust the productivity index in the same way as in equation (5.5) by multiplying the ratio of the biomass, or stocks, in the two periods of interest.

In the case of indexes involving more than two periods the choice of price weights and comparisons across periods becomes a little more complicated. We could calculate a Laspeyres quantity index between period t and $t+1$ by using the prices in period t to weight or aggregate the quantities in the beginning and last period. Similarly, we could calculate a Laspeyres quantity index between period $t+1$ and $t+2$ using the same price weights. Such direct comparisons between two periods, however, can soon become cumbersome if we are calculating quantity indexes over many periods because, if there are T periods, the total number of comparisons between any two periods is equal to $\frac{T(T-1)}{2}$. Thus if there are 10 periods we would need to calculate 45 direct indexes to ensure every possible pair of years are compared. As a result most comparisons using indexes over time are usually only calculated for each adjacent period such that a comparison is made between a given period and the immediately preceding period of interest, for a total of $T-1$ direct comparisons. For example, in the three period case presented in Table 5.1 there would be only two direct indexes to calculate: a comparison between periods t and $t+1$ and a second comparison between periods $t+1$ and $t+2$.

When comparing adjacent periods the weights used to aggregate outputs or inputs can change every year, or can be constant and equal to the base period prices (period t) in the case of a Laspeyres quantity index, or fixed at the most recent period (period $t+2$) in the case of a Paasche quantity index. The advantage of using fixed weights over time is that it allows for direct comparability across periods because all quantities have been aggregated using the same prices or weights. The principal disadvantage is that fixing the weights to a particular period (base or current period) is that as the mix of species or input combinations change with the passage of time, the prices used with a fixed base may not adequately represent the outputs and inputs produced or used by fishers, thereby making both aggregation and meaningful comparisons more difficult.

A way to overcome the potential problem of using a fixed base, especially with a long time series, is to change the base or weights that we use to make the most use of price and quantity information available to us. This could be done periodically (say every five years), or it could be done every year. For example, if we changed the base every year then the Laspeyres quantity index we calculate would use the prices as weights that were observed in the immediate preceding period, rather than the prices in the base period. To show how the choice of the price weights can affect the value of an index, a Laspeyres quantity index between period $t+1$ and $t+2$, using fixed base (t) period prices, is calculated as follows:

$$\frac{\sum_{i=x}^{z} p_{it} \times q_{it+2}}{\sum_{i=x}^{z} p_{it} \times q_{it+1}} = \frac{\$10\times3+\$5\times2+\$6\times4}{\$10\times2+\$5\times1+\$6\times2} = \frac{64}{37} = 1.7297$$

By contrast, a Laspeyres quantity index between period $t+1$ and $t+2$, but using the $t+1$ period rather than base (t) prices, is calculated as follows:

$$\frac{\sum_{i=x}^{z} p_{it+1} \times q_{it+2}}{\sum_{i=x}^{z} p_{it+1} \times q_{it+1}} = \frac{\$8\times3+\$9\times2+\$7\times4}{\$8\times2+\$9\times1+\$7\times2} = \frac{70}{39} = 1.7949$$

The difference between the two quantity indexes is explained by differences in prices period t and period $t+1$ because the quantities in the numerator and denominator are identical.

If the price weights change every year it is also possible to link or 'chain' the indexes with each other. Such a chain represents a cumulative index because it not only depends on changes in prices and quantities between the two periods, but also on changes in the intervening period. A chained index is most useful when prices increase or decrease monotonically over time and there is a long time series. It is least preferred when prices fluctuate a lot between the base and current period, but relative prices are unchanged over the entire period, and also when the time series is very short.

A chain index for a Laspeyres, Paasche or Fisher quantity index is the multiplicative sum of the $T-1$ individual indexes calculated for each adjacent period where the price weights *change* every year. Thus a chain index for a Laspeyres quantity index between period t and period $t+2$ is calculated as follows:

$$\text{Laspeyres Qty Chain Index}_{t,t+2} = \frac{\sum_{i=x}^{z} p_{it} \times q_{it+1}}{\sum_{i=x}^{z} p_{it} \times q_{it}} \times \frac{\sum_{i=x}^{z} p_{it+1} \times q_{it+2}}{\sum_{i=x}^{z} p_{it+1} \times q_{it+1}}$$

$$= 0.9736 \times 1.7949$$

$$= 1.7475 \tag{5.9}$$

When defining a chain index, and also with many fixed base indexes, it is customary to assign a value of 1.0 or 100 to the base period. Table 5.2 illustrates the difference between a Laspeyres quantity index with fixed weights equal to base period prices, a Laspeyres quantity index that involves pair wise comparisons between adjacent periods and price weights from the immediate preceding period, and a Laspeyres quantity chain index as calculated in equation (5.9).

Table 5.2 Laspeyres Pair Wise (Fixed and Changing Base) and Chain Quantity Indexes

Given Period	Laspeyres Qty. Index (fixed base)	Laspeyres Qty. Index (changing base)	Laspeyres Chain Qty. Index between given period and base period (t)
t			1.0
$t+1$	0.9736	0.9736	0.9736
$t+2$	1.6842	1.7949	1.7475

A chain index and indexes with fixed or changing weights are all useful although they will, in general, generate different index values. It is also possible to calculate what are called multilateral indexes that can be made *transitive* such that if vessel *a* is 10 per cent more productive than vessel *b*, while vessel *b* is 30 per cent more productive than vessel *c*, it should also be the case that in a pair-wise comparison between vessels *a* and *c* that *a* is 43 per cent more productive than vessel *c*, i.e., $1.1 \times 1.3 = 1.43$. Transformation of non-transitive indexes into a transitive index can be accomplished for all the indexes discussed above using what is called the EKS method (named after Elteto-Koves and Szuc), but this requires using all the observations in the sample and the calculated indexes are also subject to other weaknesses (Hill 2004). The key is to use these indexes to provide meaningful measures of changes in outputs and inputs, and hence productivity, over time and across vessels so as assess, analyze and improve the economic performance of fisheries.

5.4 'Decomposing' Profits and Measuring Productivity

Explaining productivity performance, or understanding the causes of declines or increases in productivity, is as important as measuring it. Fortunately there is an easy-to-apply method that 'decomposes' changes in relative profit performance into differences in output prices and input prices, adjusted for their importance in the catch (outputs) and fishing effort (variable inputs), and fixed inputs, such as vessel size. The approach is called an *index-number profit decomposition* and allows fishery regulators to assess changes in economic performance across vessels and over time while explicitly accounting for changes in prices and fish stocks.

For any productivity index used to compare different vessels, a decision needs to be made as to which base or numéraire should be used for comparison, and what weights should be used when aggregating multiple outputs and inputs. A useful choice of a base or reference vessel is that which is the most profitable. In this way the analysis immediately leads us to ask why the reference vessel is so profitable, and what other vessels could do to improve their profitability.

The first step in calculating the profit decomposition is to calculate the profits of all vessels and adjust them for differences in the size of fish stocks or abundance over different periods. In this way we can determine the stock-adjusted profits and, hence, pick as our reference vessel that which has the greatest profit per unit of the stock. One way to do this is to divide vessel profits by the size of the fish stock at each point in time. Another approach is to account for changes in both the total allowable catch (TAC) and the size of the fish stock to calculate a stock index (Stock^t) that is the available biomass per unit of the allowable harvest at time period t, i.e.,

$$\text{Stock}^t \equiv \frac{\text{Biomass}^t}{\text{TAC}^t} \tag{5.10}$$

Thus, for a given TAC, an increase in the biomass should make it easier to catch the regulated TAC and, all else equal, raise profits. Consequently, a stock-adjusted profit index between an arbitrary vessel b and a reference vessel a that maximizes stock-adjusted profits is given by equation (5.11).

$$\text{Stock-adjusted Profit Index}^{a,b} = \theta_s^{a,b} \equiv \frac{\pi^b}{\text{Stock}^b} / \frac{\pi^a}{\text{Stock}^a}$$

$$\equiv \frac{\pi^b}{\pi^a} \times \frac{\text{Stock}^a}{\text{Stock}^b} \tag{5.11}$$

where π^b is the profit of an arbitrary vessel b, π^a is the profit of the reference vessel a, Stock^b is the relevant stock for vessel b and Stock^a is the relevant stock for vessel a, defined by equation (5.10). If vessels a and b are observed in the same period, or if the ratio of the biomass to the TAC is identical, but profits are measured in different periods, then Stock^a = Stock^b.

The next step in calculating the profit decomposition is to decompose differences in the profit index into contributions from prices, fixed inputs and stock-adjusted productivity. This is best accomplished by defining inputs as if they were negative outputs such that variable profits are defined as the sum of all output and input (treated as negative output) quantities multiplied by their respective prices. Thus if there is just one output (q_1) and its price is p_1, and only one input (q_2) and its price is p_2, then the profit index between vessels b and a can be calculated as in equation (5.12).

$$\text{Profit Index}^{a,b} = \theta^{a,b} = \frac{\pi^b}{\pi^a} = \frac{(p_1^b \times q_1^b) + (p_2^b \times -q_2^b)}{(p_1^a \times q_1^a) + (p_2^a \times -q_2^a)} \tag{5.12}$$

Similarly, if we could construct a suitable price index (including both output and input prices) for the two vessels, defined as $P^{a,b}$, we could derive an 'implicit' quantity index which is the value difference (measured by the profit ratios) divided by the price difference (measured by the price index), i.e.,

$$Q^{a,b} = \frac{\theta^{a,b}}{P^{a,b}} \tag{5.13}$$

The implicit quantity index $Q^{a,b}$ is a measure of relative production of the two vessels, and if divided by a fixed inputs quantity index, defined by $K^{a,b}$, gives us a productivity index between vessel a and b, i.e.,

$$R^{a,b} = \frac{Q^{a,b}}{K^{a,b}} = \frac{\dfrac{\theta^{a,b}}{P^{a,b}}}{K^{a,b}} \tag{5.14}$$

A stock-adjusted productivity index can also be calculated by multiplying (5.14) by the ratio of stocks observed for vessels a and b, i.e.,

$$R_s^{a,b} = R^{a,b} \times \frac{\text{Stock}^a}{\text{Stock}^b} \tag{5.15}$$

Equation (5.15) is an index of productivity because it represents differences in the implicit quantity index that *cannot* be explained by differences in the fixed input index or fish stocks. Rearranging equation (5.14) and also adjusting for differences in fish stocks provides us with an index number profit decomposition between vessels *a* and *b*, i.e.,

$$\theta_s^{a,b} = R_s^{a,b} \times P^{a,b} \times K^{a,b} \tag{5.16}$$

To calculate the profit decomposition we need an index of prices between vessels, and also a fixed input index. A type of index that has a number of desirable properties that can be used to construct both a price and fixed input index is the Törnqvist Index (Törnqvist 1936). The reason why we need to create a price index is because we need to 'weight' output and inputs in an appropriate way that reflects their relative contributions to profits. For example, it would be inappropriate to give the price of an output that contributed only 1 per cent of a vessel's total revenue the same weight as an output that contributed 99 per cent of the vessel's total revenue.

To provide the appropriate weights the Törnqvist index uses the output and input (treated as negative outputs) shares to variable profit for each output and input, i.e.,

$$P^{a,b} = p_1^{a,b} \times p_2^{a,b} \times ... p_N^{a,b} \tag{5.17}$$

where the overall price index between the two vessels is itself decomposed into separate price indexes for each output and input denoted by *n*, and defined by their respective shares of variable profit for each vessel, i.e.,

$$p_n^{a,b} = \left(\frac{p_n^b}{p_n^a} \right)^{(s_n^b + s_n^a)/2} \tag{5.18}$$

The profit shares or weights in the exponent term of equation (5.18) are defined as $s_n^b = \frac{p_n^b \times q_n^b}{\pi^b}$ and $s_n^a = \frac{p_n^a \times q_n^a}{\pi^a}$ where if the price refers to an output then the quantity is treated as a positive number, and if the price refers to an input then the quantity used is treated as a negative.

If there are multiple fixed inputs that characterize vessels then a similar exercise is required to calculate $K^{a,b}$, i.e.,

$$K^{a,b} = K_1^{a,b} \times K_2^{a,b} \times ... K_M^{a,b} \tag{5.19}$$

where the overall fixed input index between the two vessels can be decomposed into separate fixed input indexes denoted by m with shares defined as the proportion of variable profit attributable to each fixed input. Calculating such shares requires some method to allocate variable profits among the fixed inputs. If there is only one fixed input (such as vessel size) and assuming that all the variable profit can be attributed to this fixed input, then equation (5.19) simplifies to the ratio of the fixed input between any arbitrary vessel and the reference vessel, i.e.,

$$K^{a,b} = \frac{k^b}{k^a}$$

(5.20)

where k^b is the fixed input measure for vessel b and k^a is the fixed input measure for vessel a.

Equations (5.15) to (5.20) can be calculated using a spreadsheet program such as Excel, or other software. To show how these calculations are made, Table 5.3 provides a hypothetical example of two vessels each with one output, one variable input and one fixed input.

Table 5.3 Hypothetical Prices and Quantities for Two Vessels

Variable	Vessel a	Vessel b
Output Price (p_1)	$10	$8
Output Quantity (q_1)	7	5
Variable Input Price (p_2)	$4	$3
Variable Input Quantity (q_2)	3	2
Fixed Input Quantity (k)	20	10
Fish Stock Size (s)	16	14

The first step to derive the index number profit decomposition is to calculate the variable profits, defined as total revenue less variable costs, i.e.,

Variable Profit Vessel a = π^a = ($10X7)-($4X3) = $58
Variable Profit Vessel b = π^b = ($8X5)-($3X2)   = $34

The variable profits, however, need to be adjusted for differences in the fish stock by dividing the variable profits by the relevant stock size. Thus the stock-adjusted variable profits are calculated as follows:

$$\text{Stock-adjusted Variable Profit Vessel } a = \frac{\pi^a}{s^a} = 3.63$$

$$\text{Stock-adjusted Variable Profit Vessel } b = \frac{\pi^b}{s^b} = 2.43$$

Given that vessel a has the largest stock-adjusted variable profit we treat it as the reference or benchmark vessel that acts as the denominator in all the indexes. The profit index (with and without the stock adjustment) between the two vessels is, therefore,

$$\theta^{a,b} = \frac{\pi^b}{\pi^a} = \frac{\$34}{\$58} = 0.59$$

$$\theta_s^{a,b} = \frac{\pi^b}{\pi^a} \times \frac{\text{Stock}^a}{\text{Stock}^b} = 0.59 \times 1.14 = 0.67^1$$

In the stock-adjusted case, part of the difference in profitability between the two vessels is attributed to differences in the size of the fish stock when the vessels were observed. Because vessel a fished with a larger fish stock (16 units for vessel a and 14 units for vessel b) the stock-adjusted profit index is higher than the unadjusted profit index thus indicating a relatively better performance for vessel b after accounting for differences in fish stocks.

Using the Törnqvist price index specified in equation (5.17) we calculate a price index for both vessels that we can then use to derive the implicit quantity index, i.e.,

$$P^{a,b} = \left(\frac{p_1^b}{p_1^a}\right)^{(s_1^b + s_1^a)/2} \times \left(\frac{p_2^b}{p_2^a}\right)^{(s_2^b + s_2^a)/2} = \left(\frac{8}{10}\right)^{(s_1^b + s_1^a)/2} \times \left(\frac{3}{4}\right)^{(s_2^b + s_2^a)/2} = 0.81$$

where

$$s_1^b = \frac{8 \times 5}{34} = 1.18, \quad s_1^a = \frac{10 \times 7}{58} = 1.21, \quad s_2^b = \frac{3 \times -2}{34} = -0.18 \quad \text{and} \quad s_2^a = \frac{4 \times -3}{58} = -0.21$$

---

1   Without rounding to the second significant decimal place the stock-adjusted profit index is 0.6699. However, rounding the profit ratio (0.59) and the ratio of the stocks (1.14) to the second decimal point and then multiplying their product is 0.6726.

The implicit quantity index is calculated using the profit index and the price index as follows,

$$Q^{a,b} = \frac{\theta^{a,b}}{P^{a,b}} = \frac{0.59}{0.81} = 0.72$$

The productivity index is simply the ratio of the implicit quantity index over the fixed input index. As there is only one fixed input there is no aggregation problem, and the fixed input index is simply the ratio of the size of vessel b and vessel a, i.e.,

$$K^{a,b} = \frac{k^b}{k^a} = \frac{10}{20}$$

The productivity index is, therefore,[2]

$$R^{a,b} = \frac{Q^{a,b}}{K^{a,b}} = 1.45.$$

The stock-adjusted productivity index is

$$R_s^{a,b} = R^{a,b} \times \frac{\text{Stock}^a}{\text{Stock}^b} = 1.45 \times 1.14 = 1.66$$

The full profit decomposition after accounting for differences in fish stocks between vessels can, therefore, be written as follows,

$$\theta_s^{a,b} = P^{a,b} \times K^{a,b} \times R_s^{a,b} = 0.81 \times 0.5 \times 1.66 = 0.67$$

where the price index can be decomposed into the price index for the output and the price index for the input, i.e.,

$$P^{a,b} = p_1^{a,b} \times p_2^{a,b} = 0.77 \times 1.06 = 0.81$$

---

2   If we calculate the implicit quantity index to the fourth significant decimal place and then divide it by the fixed input index (0.5) we obtain a value for the productivity index of 1.45 when rounded to the second significant decimal place, rather than $1.44 = 0.72/0.5$.

When interpreting the index-number profit decomposition it should be emphasized that the calculated indexes are all relative to a base or reference vessel. Thus if a different vessel is chosen for the reference vessel, a different set of indexes will be obtained. The choice of the reference vessel is especially important in terms of the productivity index because the larger is the reference vessel for a given profit level, the larger will be the productivity index of any arbitrary vessel (vessel b in our example).

A value *greater than one* for any of the components ($p_1^{a,b}$, $p_2^{a,b}$ and $R_s^{a,b}$), but not $K^{a,b}$, implies that the contribution of the respective index to differences in variable profits is greater for the vessel in question (vessel b in our example) than it is for the reference vessel. By contrast, the fixed input quantity index simply records relative differences in fixed inputs that may, or may not, contribute to differences in variable profits. Moreover, differences in the vintage and quality of vessels, which are part of the multiple characteristics of vessels, will often be obscured when using a simple one characteristic measure, such as vessel size. In the case of the output price index ($p_1^{a,b}$ in our example) the value 0.77 is less than 1.0 which implies that part of the reason why profits are greater for vessel a relative to vessel b is the higher price (vessel b receives \$8/unit while vessel a receives \$10/unit) obtained by the reference vessel for its harvest. In terms of the input price index ($p_2^{a,b}$ in our example) the value of 1.06 implies that its price (vessel b pays \$3/unit while vessel a pays \$4/unit) is such that it gives vessel b a profit advantage relative to vessel a. If the prices of the input or the output are exactly the same for the arbitrary and reference vessel then the price indexes will equal 1.0. Apart from the price it pays for its input, the other advantage vessel b has over vessel a is a higher level of productivity, as indicated by a stock-adjusted productivity index of 1.66. The productivity is higher for vessel b, despite the fact its profits are lower, because it is able to generate a higher profit per unit of the fixed input than does vessel a after accounting for differences in fish stocks and also output and input prices.

5.5 Fisheries Applications

Productivity measures and index number decompositions can be used in many different ways in fisheries. The first application of the index number profit decomposition was in the British Columbia (BC) halibut fishery. This fishery is of interest because it switched from the exclusive use of input controls to individual transferable quotas in 1991. Using data on vessels for 1988, 1991 and 1994 it is possible to measure changes in productivity, and also decompose differences in profits relative to a reference vessel.

The cells in Table 5.4 represent the geometric means of the different indexes and the arithmetic mean of the actual profits by year and vessel class in the BC halibut fishery. The profit index, defined relative to the most profitable vessel that is observed in 1994, is available as both an unadjusted (θ) and a stock-adjusted value (θ_s). The stock-adjusted value is identical for all vessels observed in 1994 because the reference vessel is also observed in 1994. The different components of the price index are defined as output price of halibut (*PH*), price of fuel (*PF*) and the price of labor (*PL*) and are obtained from real prices (adjusted for inflation). The fixed input index (*K*) is defined as the ratio of vessel length relative to the reference vessel. Productivity (R_s) represents differences in profit per unit of vessel length after accounting for price differences and fish stocks.

Table 5.4 Profit Decomposition in the BC Halibut Fishery (1988-1991)

Obs.	Number	Profit($)	θ_s	R_s	PH	PF	PL	K
1988	43	103,898	0.242	0.706	0.429	0.995	1.007	0.795
Small	24	70,298	0.162	0.593	0.412	0.994	1.008	0.661
Large	19	142,539	0.386	0.870	0.450	0.996	1.006	0.983
1991	43	48,851	0.068	0.147	0.667	1.004	1.000	0.688
Small	33	34,482	0.061	0.141	0.659	1.004	1.000	0.637
Large	10	83,369	0.139	0.218	0.700	1.003	1.001	0.904
1994	19	126,521	0.331	0.486	0.844	1.000	0.998	0.808
Small	9	66,079	0.184	0.346	0.806	1.000	0.997	0.662
Large	10	180,919	0.561	0.660	0.880	1.000	1.000	0.967

Source: Fox et al. (2003, p. 171) and authors' calculations.

The profit decomposition provides an explanation as to why average profits are, on average, the highest in 1994. The output price index (*PH*) shows increases in prices from 1988 to 1991, and also from 1991 to 1994, have helped to increase the profitability of vessels over time. The reason for the large and positive change in *PH* lies in the switch to the use of individual vessel output controls in 1991 that made the restrictions on season length redundant. As a result, the fishing season increased from 6 days in 1990 to 214 days in 1991, and eventually 245 days in 1994. This allowed fishers to take their time when catching halibut and increase quality, and also to land much of their harvest as fresh product which commands a much higher price than if frozen.

The other important change over the period is the initial decline in the productivity index from 1988 to 1991, and its subsequent increase from 1991 to 1994. A possible reason for the decline to 1991 is that the almost 50 per cent reduction in the TAC between 1988 and 1991 constrained the profits of fishers by limiting their harvests. Moreover, the much extended fishing season in 1991 led fishers to change their previous catch times and locations which may, initially, have lowered productivity as they experimented how best to optimize their operations. Only in 1993 when transferability of the individual harvesting rights was permitted were fishers able to overcome this constraint and, thus, increase their productivity.

Another example illustrates the ability of the decomposition to analyze multiple output price effects on relative profits. Using vessel data available for 1988, 1990 and 1991 from Canada's multispecies mobile gear fishery in Nova Scotia and New Brunswick, the profit decomposition is able to discern changes brought about by the introduction of individual transferable quotas for three species (cod, haddock and pollock) in 1991.

To compare the impact of the quotas across species, a profit decomposition can be calculated for the 10 major species caught by fishers: the three quota species and the six most important species caught without quota, plus a category for all other fish species. The geometric means of the price indexes for all these species is presented in Table 5.5. The results suggest that implementing ITQs for cod, haddock and pollock had a beneficial impact on prices in 1991 that translated into higher prices, and thus profits, for vessels.

Table 5.5 Geometric Means of Output Price Indexes by Species and Year in Scotia-Fundy Mobile Gear Fishery

Variable	1988	1990	1991
Quota Species	1.147	1.381	1.995
Price Index Cod	1.088	1.171	1.453
Price Index Haddock	1.045	1.087	1.228
Price Index Pollock	1.009	1.086	1.118
Non-quota Species	0.970	0.971	0.996
Price Index Redfish	0.995	0.996	0.992
Price Index Flounder	0.977	0.979	1.002
Price Index Halibut	1.000	0.997	1.001
Price Index Hake	1.000	1.003	1.002
Price Index Cusk	1.000	1.000	1.000
Price Index Catfish	0.997	0.996	0.999
Price Index All Others	1.001	1.000	1.001

Source: Dupont et al. (2005, p. 49).

Another interesting analysis is to compare productivity changes over the period 1988 to 1991. Dividing the Nova Scotia-Fundy mobile gear fleet into those vessels less than 45 feet, and equal to a greater than 45 feet in length, we can investigate the trend in productivity for small and large vessels. This is summarized in Table 5.6 where θ is the profit index, R is the productivity index, PO is the overall output price index, PF is the fuel price index, PL is the labor price index and K is the ratio of vessel size to the reference vessel. The numbers are not stock-adjusted as there is insufficient data on biomasses to make the appropriate calculations.

Table 5.6 Geometric Means of Profit Decomposition in Scotia-Fundy Mobile Gear Fishery

Vessel Type	θ	R	PO	PF	PL	K
1988						
Small	0.272	0.248	1.148	1.000	2.176	0.439
Large	0.474	0.271	1.073	1.000	1.795	0.908
1990						
Small	0.226	0.167	1.441	0.987	2.368	0.401
Large	0.359	0.145	1.284	0.987	2.069	0.946
1991						
Small	0.187	0.107	1.975	0.963	2.565	0.357
Large	0.564	0.207	2.001	0.972	1.461	0.960

Source: Dupont et al. (2005, p. 51).

The results given in Table 5.6 show a decline in productivity for both vessel sizes from 1988 to 1990 that coincided with substantial declines in the groundfish stocks. However, there was an increase in productivity in 1991 for large vessels despite very substantial declines in the biomass of the major species from 1990 to 1991. The results suggest that large vessels in this fishery were better able to adjust and benefit from productivity improvements following the introduction of individual transferable quotas in 1991 for cod, haddock and pollock. Both vessel classes, however, benefited substantially from a rise in output prices for the three quota species, but the increase was not sufficient for small vessels overall to overcome declines in productivity from 1990 to 1991 which reduced their relative profit performance from 1988 to 1991.

A final example of the profit decomposition method can be used to assess the effects of a vessel buyback in the south-east trawl fishery off the east coast of Australia. Table 5.7 provides a summary of the results by year in terms of the geometric means of all the individual vessel indexes, with the column headings defined identically as in Table 5.6. The key result is that it seems a vessel buyback instituted in 1997, coupled with improvements in quota trading, helped to raise productivity of both small and large vessels in 1998 and may have contributed to further increases in productivity of small vessels in 1999 and 2000.

Unlike vessel or license buybacks implemented in other fisheries, such as British Columbia's salmon fishery or the US northeast multi-species fisheries (Holland et al. 1999), these productivity gains have occurred within a fishery managed by individual and transferable output controls. The profit decomposition suggests that the buyback, combined with individual tradeable harvesting rights and greater quota trading through the establishment of a quota brokerage service, have been successful at improving economic performance over the period 1997-2000 with a rise in the mean profit index for both vessel classes.

Table 5.7 Profit Decomposition in the South-East Trawl Fishery of Australia

Obs	*No.*	*Profit*	q_s	R_s	*PO*	*PF*	*PL*	*K*
1997	30	176,336	0.055	0.153	0.199	1.057	5.643	0.303
Small	19	90,327	0.037	0.146	0.187	1.049	6.368	0.203
Large	11	324,896	0.108	0.166	0.221	1.073	4.581	0.602
1998	33	203,622	0.074	0.233	0.238	1.068	4.117	0.306
Small	20	99,080	0.047	0.231	0.221	1.059	4.435	0.195
Large	13	364,457	0.151	0.238	0.263	1.082	3.670	0.608
1999	29	282,937	0.113	0.247	0.374	1.047	3.105	0.337
Small	17	159,367	0.083	0.376	0.359	1.034	2.913	0.204
Large	12	289,665	0.174	0.178	0.395	1.064	3.398	0.686
2000	28	276,171	0.117	0.297	0.360	1.000	3.313	0.331
Small	17	149,369	0.081	0.390	0.336	1.000	3.078	0.202
Large	11	472,139	0.205	0.195	0.399	1.000	3.713	0.709

Source: Fox et al. (in press) and authors' calculations.

5.6 Conclusions

The measurement and analysis of productivity is an important area of study in economics, and is of particular importance in fisheries management. By tracking changes in fleet productivity over time managers are able to identify changing economic conditions, and can separate these changes from variations in fish stocks or abundance. Comparisons across vessels of productivity, especially in terms of profit decompositions, provide a way to improve profits and productivity by determining what factors are constraining vessel performance. Productivity indexes can be easily calculated using spreadsheet programs, and also have the advantage that they impose no restrictions on the technology of vessels, nor do they require the assumption of profit maximization.

References

Dupont, D.P., Fox, K.J., Gordon, D.V., and Grafton, R.Q., 'Profit and Price Effects of Multi-species Individual Transferable Quotas', *Journal of Agricultural Economics*, 56 (2005): 31-57.

Fisher, I., *The Making of Index Numbers* (Boston: Houghton, Mifflin, 1922).

Fox, K.J., Grafton, R.Q., Kirkley, J., and Squires, D., 'Property Rights in a Fishery: Regulatory Change and Firm Performance', *Journal of Environmental Economics and Management*, 46 (2003): 156-77.

Fox, K.J., Grafton, R.Q., Kompas, T., and Che, T.N., 'Capacity Reduction, Quota trading and Productivity: The Case of a Fishery', *Australian Journal of Agricultural and Resource Economics* (in press).

Hill, R.J., 'Constructing Price Indexes Across Space and Time: The Case of the European Union', *American Economic Review*, 94/5 (2004): 1379-410.

Holland, D., Gudmundsson, E., and Gates, J., 'Do Fishing Vessel Buyback Programs Work: A Survey of the Evidence', *Marine Policy*, 23/1 (1999): 47-69.

Squires, D., 'Productivity Growth in Common Property Resource Industries: An Application to the Pacific Coast Trawl Industry', *RAND Journal of Economics*, 23/2 (1992): 221-36.

Squires, D., 'Sources of Growth in Marine Fishing Industries', *Marine Policy*, 18/1 (1994): 5-18.

Törnqvist, L., 'The Bank of Finland's Consumption Price Index', *Bank Finland Monthly Bulletin*, 10 (1936): 1-8.

Chapter 6

Economics for Fisheries Management

> While any renewable resource poses difficult management problems, marine fisheries are especially hard to manage.
>
> *Jared Diamond*
> *Collapse: How Societies Choose or Fail to Survive (2005, pp. 479-480)*

6.1 Introduction

Fisheries managers face the difficult task of ensuring the sustainable use of a highly variable resource while maintaining healthy marine ecosystems. Not surprisingly, they have not always made the best decisions while trying to juggle the many demands asked of their time and expertise. In this, our concluding chapter, we briefly review the various fisheries applications we have presented previously, and summarize what they imply about the economics of fisheries management. We also provide our perspective on the problems with input controls, instrument choice in fisheries, the challenges of uncertainty, the value of adaptive management of fisheries, and what is required to improve fisheries performance.

6.2 Economic Insights for Fisheries Management

The eight different fisheries discussed in the book are located in four different parts of the world (western and eastern Canada, eastern USA, west and east coast Malaysia and northern and south-eastern Australia) and include artisan fisheries (Malaysian gillnet fishery), fisheries where the harvesters explicitly target multiple demersal species (Nova Scotia mobile gear fleet, Australia's south-east trawl fishery), pelagic species (Australia's northern prawn fishery), and fisheries managed with input controls (Australia's northern prawn fishery) and individual output controls (British Columbia halibut fishery, Nova Scotia mobile gear fishery, Australia's south-east trawl fishery). Two important generalizations from these case studies are worth highlighting. First, there are large variations in economic performance among fishers and second, the potential exists to substantially improve the economic performance of fishers. In this chapter we present our perspectives on how such improvements can be achieved while also maintaining the sustainability of fisheries.

One of the key insights of this book is that fisheries differ a great deal and there can be no 'one-size-fits-all' approach to fisheries management. Nevertheless, experiences from many jurisdictions suggest that there are some generalizations that can guide managers who wish to ensure sustainable and economically viable fisheries. We summarize these insights as six 'rules of thumb' regarding the economics of fisheries management:

1. Regulations (especially input controls) affect fisher behavior and performance, often in negative and counterproductive ways that can be damaging to fisheries. For example, decreasing the fishing season to protect stocks and to ensure the total allowable catch is not exceeded can exacerbate the excess capacity problem because fishers have an even greater incentive to invest in inputs to catch their harvest in a shorter period of time.

2. Management regulations that run counter to the incentives faced by fishers will, in general, be much less successful than those that allow fishers to respond to signals that allow their private interests to coincide with societal interests. Fishers, as do managers, respond to incentives and the best way to get people to do something is to make sure that it is in their interest to do it.

3. Fisheries managers simply do not have the information to adequately manage fisheries through a system of top-down controls. The vagaries of stock assessment and the inability to fully observe actions of fishers at sea means that detailed regulations and controls are unlikely to optimize management actions. Instead, fishers themselves – if given the right incentives – are often the best people to work out how to address stock or other problems at a local level.

4. Management responsibilities and secure harvesting rights encourage sustainable behavior by fishers. Property rights provide fishers with a long-term incentive to protect stocks and habitat while responsibilities promote collective action. Secure property rights also engender social capital within fishing communities that has the potential to reduce management costs and improve fisheries outcomes (Grafton, 2005).

5. In many (but not all) fisheries there is no trade-off between biological and economic objectives in terms of stock size. Provided that harvesting costs increase the smaller the fish stock, and even with high discount rates, the maximum economic yield (see chapter 1) is almost always at a greater stock size than that which maximizes maximum sustainable yield.

6. Fisheries management that promotes long-term economic viability of fishing also supports sustainable fisheries. In other words, good economic management of fisheries complements approaches that are designed to ensure the long-term

sustainability of marine resources. Fisheries that are characterized by chronic excess capacity and poor economic returns are almost invariably those that are overexploited biologically, and are less resilient to negative environmental shocks.

Changing current management structures from 'command and control' to incentives and responsibility-based management is not an easy task for either fishers or managers, but it does offer substantial rewards in terms of sustainability (Grafton et al. 2006; Hilborn et al. 2005), improved economic outcomes and the potential to reduce management costs (McLoughlin and Findlay 2005).

6.3 Problems with Input Controls

The general lack of success of input management regimes (Townsend 1990) arises from two causes. First, controls on one or more inputs provide an incentive for operators to substitute uncontrolled inputs so as to achieve their desired catch. Consequently, a manager relying on input controls is in constant competition with the imagination, energy and inventiveness of each operator in the fishery and the full technological backup of a modern economy. Effort creep is inevitable. Second, input control regimes provide very little, if any, sense of ownership or stewardship of the fisheries resource. There are no guarantees in any input control management regime, except the right of access to the fishery under certain guidelines. Operators are encouraged by these rules to compete for catch within those rules, and if one operator refuses to expand effort, while others do, that operator will be worse off.

The problem of inputs controls is epitomized by the recent history of the Australian Northern Prawn Fishery (NPF). Over the past thirty years the NPF has been managed by a series of input controls, including seasonal closures, a move from quad to twin nets, engine power and hull limits and, most recently, gear reductions and restrictions. In all cases, the limits to fishing power have been temporary, at best. For example, limits placed on vessel A-units (a measure of hull capacity and engine power) led to substitution toward unregulated inputs, specifically gear headrope length. Thus, as a result of controls on A-units, fishers have responded by increasing the average headrope length used on their vessels from 23.12m in 1993 to 26.06m in 2000 (Kompas et al. 2004).

The implication of this counter-movement in A-units and headrope length is twofold. First, restricting A-units has failed to control effort because fishers have successfully substituted to other inputs, and used existing inputs more intensively. Second, the forced change in input combinations, inducing boat owners to use different proportions of gear to A-units, has resulted in considerable loss in boat-level efficiency, as reported in chapter 3.

Under the system of input controls, every change made in the management regime in the NPF – seasonal and area closures, A-unit restrictions and most recently gear reductions – was made in recognition that the system it replaced had failed to sufficiently constrain effective effort and the inevitable effort creep. Increased fishing power along with declining catches has made it increasingly clear that prawn stocks are not being conserved and catches and effort are not being controlled (Rose and Kompas 2004).

6.4 Policy Choices and Fisheries Management

In 2002 and 2003 the Food and Agricultural Organization of the United Nations (FAO) sponsored two workshops on factors that contribute to overexploitation and unsustainability in fisheries. The conclusion of the experts who attended the meetings was that six factors contribute to unsustainability: one, inappropriate incentives (especially those that contribute to overcapitalization and the race to fish), two, high demand for fish, three, poverty and lack of alternatives for fishers, four, complexity and inadequate knowledge, five, ineffective governance (especially conflicting objectives and inability to implement effective management measures) and, six, failure to adequately consider interactions between fishing and other sectors, and also the environment (FAO 2004). The economics of fisheries management plays an important role in the approaches used to resolve or mitigate these factors. In particular, economics can assist greatly in three of the eight approaches proposed by the FAO group of experts to address unsustainability that include the granting of secure rights to fishers, using economic tools to understand the distribution of benefits from fisheries, and the use of market incentives.

In this section we focus on the use of market incentives in the form of individual harvesting rights as a way to improve overall fisheries management. With effort creep an inevitable outcome of input controls, economists generally argue for overall catch controls combined with individual and transferable harvesting rights for fishers, commonly called individual transferable quotas (ITQs) or individual fishing quotas (IFQs). These harvesting rights, if properly enforced, ensure that each vessel is allowed to harvest a given share of the total allowable catch (TAC). Such rights mitigate, and may even entirely remove, the race to fish. This is because fishers are assured that they can catch their given share of the total catch and, thus, do not need to make investments or to unnecessarily rush their fishing to harvest their desired amount of fish before the season ends. Given adequate monitoring and enforcement ITQs have worked well for decades in fisheries throughout the world, including New Zealand, Iceland, the USA, Australia and Canada, and have also helped to maintain stocks at sustainable levels (Organisation for Economic Co-operation and Development 1997; Hannesson, 2004). In addition, they have led to a range of economic benefits, as illustrated by the British Columbia halibut fishery (see chapters 3 and 5), in terms of cost savings, improvements in efficiency and increased revenues (Grafton et al. 2000a).

The potential benefits of ITQs arise from a number of factors. First, because these rights are tradeable, market forces will generally distribute quota among fishers that value the right most highly. Vessels that have lower marginal costs of fishing will, thus, be willing to pay more for quota, with the resulting transfer of quota from high to low marginal cost producers increasing economic efficiency overall – essentially fishing inputs are distributed to those who use them best. In other cases, quota trades allow vessels to compensate for catches that are larger or smaller than their existing quota holdings. These efficiency gains (or what amount to cost reductions) can be substantial, even in fisheries where TAC is not binding in aggregate. In the Australian south-east Trawl Fishery, for example, the cost savings from quota trades are estimated to be 1.8-2.1 cents a kilogram for every 1 per cent increase in the volume of quota traded (Kompas and Che 2005).

Second, instead of investing in boat capacity to catch fish before others do, with a much greater assurance of catch with a harvesting right, vessel owners can, instead, concentrate on investments that lower the per unit costs of fishing. This is a major benefit as boat-specific technological change lowers the costs of fishing and increases profits, but does not impose costs on other fishers. Nor does it have negative implications for the stock provided the TAC is set appropriately. By contrast, with input controls, technological change – new boats, a better engine, more efficient gear, try nets, navigation gear and other innovations – is harmful in the sense that increased fishing power lowers fishery profits and endangers stocks. In some cases, regulations to control effort creep are designed to prevent the very adoption of new technologies that, under other circumstances, may be beneficial or efficiency enhancing.

A third potential benefit of individual harvesting rights is that regulations established to lower the effectiveness of fishing effort, and that are costly to implement and enforce, may be eliminated. Spawning stocks must naturally be protected and marine reserves can almost always be justified even on economic grounds (Grafton et al. 2005b), but area and seasonal closures used to simply limit effort are unnecessary where the landings (and possibly catch) of each fisher is monitored. By removing such controls, operators can fish when the weather permits, and perhaps more importantly, match the harvest throughout the year to market conditions, generating the highest price for their catch. This also helps fishers to choose the right mix of inputs, and the time and manner to fish, all of which is cost reducing and improves economic efficiency.

A final benefit of ITQs is that they allow for autonomous adjustment of the fishing fleet, with operators voluntarily able to 'cash out' by selling their quota to more profitable vessels. Indeed, if implemented correctly, individual harvesting rights can help generate the largest possible (marketable) asset value for those who have the right to fish, reflected in a high price for each unit of individual quota. Fishers are thus compensated for exiting the fishery, without the need for government intervention. This is in stark contrast to input controlled and over-capitalized fisheries where fishers often lobby heavily for government vessel-buyback schemes and may be forced to leave fishing with compulsory buybacks. More importantly, by eliminating the problem of effort creep, ITQs help prevent the long-term secular decline in average net returns common in many input-controlled fisheries.

To achieve the potential benefits of individual harvesting rights there must be adequate monitoring and enforcement. This can be costly, although there is no necessary reason for this cost to be a government responsibility. Under an ITQ system, fishers are keen to protect their secure property rights and it is not uncommon for monitoring to be at least partially funded by industry (Grafton et al. 2005a). Despite monitoring and enforcement, the possibility exists that fishers may dump lower valued fish at sea so as to maximize the value of their landings. Such activity is often illegal and is called 'highgrading'. It exists in input-controlled as well as output-controlled fisheries. Provided that highgrading can be estimated, the TAC can be matched with desired mortality so as to prevent overexploitation of stocks. Thus, although highgrading is wasteful and undesirable it need not lead to unsustainable fish stocks and can be considered a cost of management.

A challenge with ITQs, or any output control approach, is that managers face a problem in setting the TAC when abundance varies between seasons and is unknown at the beginning of the season. By setting the TAC too high the manager runs the risk that fishing pressure on stocks will be excessive should a low abundance season occur. By setting the TAC more conservatively, the manager guarantees the loss of potential profits, should the season be one of high abundance. Indeed, not only is the problem well recognized, it is often cited as a primary reason for preferring input controls.

What is not so well understood, however, is that essentially the same problem affects the setting of input controls. To set effort at the optimal level, the manager needs information on abundance, catch per unit effort, the value of catch and the cost of effort. Setting input controls too tightly leads to loss of potential profits in seasons of high abundance. Setting input controls too generously leads to excess investment and effort and too large a catch. The long-term consequences are pressure on future stocks and dissipation of potential profit.

6.5 Challenges of Uncertainty

Fisheries managers know that environmental factors make it very difficult to discern a signal from noise in stock assessment data, and unexpected shocks make it difficult to adjust management plans, often at very short notice. For example, even if the current level of harvesting were sustainable in the past, changed environmental conditions may result in the collapse of a fishery at the same, or even at a lower, harvest rate. Overlaying this environmental uncertainty are fluctuations in market conditions and uncertainties over the social and economic consequences of a lower total harvest.

Despite the inherent difficulties of managing under uncertainty the key principles are 'common sense'. They include one, the need to consider a wide number of possible strategies when deciding on an action, two, undertake actions that are robust or useful under a variety of scenarios and, where possible, that are reversible and informative, three, implement regular assessment and monitoring of actions and, four, update and change actions as necessary based on this information (Ludwig et al. 1993). It also requires, as we have stressed throughout this book, an understanding of human behavior and incentives because harvesting is a major driver of both fluctuations and declines in fisheries.

Uncertainty complicates management and has led some decision-makers to follow at least two sub-optimal strategies: one, collect ever more detailed biological information so as to obtain more precise estimators of biological models and two, institute ever greater controls on fishers in an attempt to ensure greater certainty over catches. The experiences of many fisheries suggest that such strategies may do little to address irreducible uncertainties, and in the long run may even be counterproductive. For example, more and better quality biological data should increase the precision of estimates of parameters in stock-recruitment or other sorts of biological models. This can create a false sense of security in the models without providing any real understanding about the nature of potential shocks to the system (Grafton et al. 2004a). Moreover, in the case of chaotic populations, no matter how accurate is the information it will never be good enough for predictive purposes (Grafton and Silva-Echenique 1997). Obtaining better stock assessments is also costly and may 'crowd out' the time and effort required to collect economic data that might, at the margin, generate a much higher payoff. Simply put, more and better information about fish stocks or the environment in the face of uncertainty does not necessarily ensure better fisheries outcomes.

The desire to control what appears to be controllable is understandable when facing uncertainty, but in reality managers in most fisheries cannot fully regulate fisher behavior, nor should they. If controls are placed on one set of inputs, fishers are adept at substituting to other and often less efficient inputs, as has occurred in Australia's northern prawn fishery, and elsewhere. This reduces economic viability and leads to effort creep and even greater uncertainties as the 'capacity overhang' increases. In turn, this increases the incentives for fishers to lobby for even greater access to cover the costs of their investments and to stay in business.

Another important consideration is the effects of uncertainty that can arise in terms of environmental variability, stock measurement and actual catches (Roughgarden and Smith 1996). Measurement uncertainty has the largest impact on optimal fisheries management and should lower the TAC compared to the case when the stock is measured with certainty (Sethi et al. 2005). Another important issue in terms of uncertainty is the choice of whether to use input or output controls. Which is preferred in an uncertain world depends on the objectives of management, and also on the relative size of the uncertainty in terms of the growth of the stock versus the catch per unit of effort (CPUE). The larger the uncertainty in terms of CPUE the more preferred will be the use of output controls relative to input controls (Danielsson 2005).

6.6 Adaptive Management in Fisheries

There is no easy way to address the challenges of uncertainty, but ignoring or being unprepared or unwilling to act or confront negative shocks that frequently arise in fisheries is a recipe for disaster. The key is to make the best use of all of the possible data that is available, to learn from this information and from management successes and failures, and most importantly, to ensure management is sufficiently flexible to change and to adapt to changing conditions. Such an approach to management was described by Walters and Hilborn (1976), and elaborated on in Walters and Hilborn (1978), and has been called active adaptive management.

The feedback or adaptive loop associated with adaptive management is illustrated in Figure 6.1. Active adaptive management involves one, setting of quantifiable goals, two, the setting of broad management actions, called strategies, three, the implementation of tactics or day-to-day actions consistent with the goals and strategies, four, monitoring, learning and evaluation from management actions and last, but not least, management response to information and analyses about the state of the fisheries.

Economics plays an important role in all five components. The economic viability of fisheries is intimately connected with sustainability and requires that economic, as well as biological, objectives be included and quantified in the goals of fisheries management. Strategies, such as whether to use input or individual output controls, requires an understanding of how fishers respond to economic incentives. Tactics, or actions implemented on a daily or weekly basis, should be consistent with overall goals and strategies. Economics also provides a framework to better understand these connections by explicitly considering the behavior of fishers. Finally, as we have shown throughout this book, economics provides a rich set of tools to learn from management actions and analyze performance that helps generate better management outcomes.

Management problems can arise at various stages in the 'adaptive cycle'. Some fisheries do not have quantifiable management objectives beyond what might be called 'motherhood and apple pie' statements, usually involving the word sustainability. As a result, management and evaluation of actions is problematic because there is a lack of quantifiable benchmarks to judge performance.

Although many actively managed fisheries will have strategies and tactics that are clear, at least to managers, many of these same fisheries lack an adequate system of monitoring and evaluation of management actions. Too often the focus is on the status of fish stocks as a proxy of management performance. Clearly this is important, but without following through with the logical connections as to how management may be contributing, or not, to the current stock status there can be no effective management. For example, if the management strategy is to regulate fishing with input controls alone then over time, with input substitution and effort creep, this could eventually result in a situation where the capacity output greatly exceeds the current harvest. At this point it becomes very difficult to ensure stocks are adequately protected with regulations such as a limited fishing season, and so

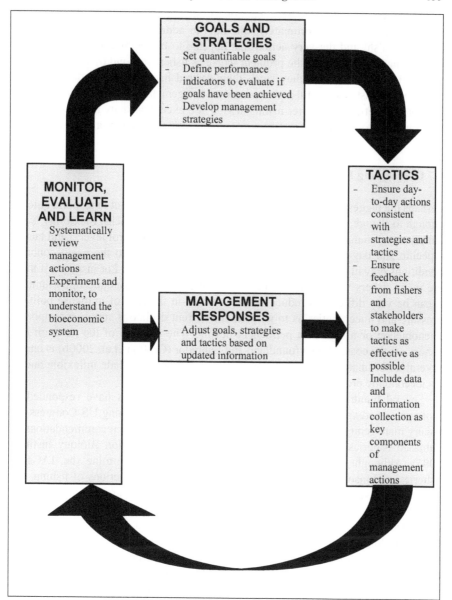

Figure 6.1 Adaptive Management in Fisheries

Source: Adapted from Grafton et al. (2004a, Figure 16.2).

managers may be tempted to be precautionary and impose a 'conservative' or a very short fishing season. Unfortunately, this fails to address the incentives of fishers to 'race to fish' and the very action that appears to be required (a shorter fishing season) may, in fact, make the problem worse as fishers respond by investing more gear and inputs to ensure they catch their desired harvest in a shorter period of time. By contrast, successful active adaptive management would identify the causal relationship between a shorter fishing season and excess capacity and, presumably, result in the adoption of alternative management strategies before a crisis occurs.

6.7 Overcoming Failures in Fisheries

Perhaps the biggest failure of fisheries management has been not to respond quickly enough to change, especially to unexpected declines in fish stocks. Flexibility is not simply about managers changing their strategies (overall control levers) and tactics (specific and day-to-day actions), but is intimately related to the socioeconomic conditions prevailing in the harvesting and processing sectors. For instance, if there are many fishers with poor incomes and with few other economic opportunities, it can be very difficult to reduce current catches in the face of immense political pressure and uncertainties as to whether the current decline is simply a temporary phenomenon or a sign of a permanent decline. The collapse of the northern cod fishery off the coast of Newfoundland and Labrador (Grafton et al. 2000b) is one of several unfortunate examples of how decision-making was made inflexible due to socioeconomic pressures.

Several hundred fisheries scientists in the United States have responded to biological overexploitation of managed fish stocks by petitioning US Congress for fishery management decisions to be based on '…the scientific recommendations of unbiased fisheries and marine scientists' (Marine Conservation Biology Institute 2005). Although science (and also economics) should determine the TACs of fisheries, and also consider environmental impacts and interactions of fishing, the reality is that without the appropriate incentives in the form of property rights fishers often have little incentive to allow depleted stocks to rebuild. As a result, with input controlled fisheries that predominate in the United States, and elsewhere, fishers frequently lobby against cuts in the total harvest. By contrast, in fisheries where harvesters have a long-term and secure right to catch a given share of the total catch they have lobbied for reductions in the TACs so as to protect their property rights and future livelihood (Grafton et al. 2006). Thus, the insight from the overexploitation of fish stocks is not to blame fishers who are simply responding to perverse incentives, but to ensure fisheries are economically viable and that fishers have the appropriate incentives and responsibilities.

A key way to improve the economic viability of fisheries is to adopt incentive based approaches such as individual harvesting rights, territorial user rights and also to support cooperative behaviour by fishers. In all these approaches the costs imposed on others from harvesting are, to a much greater extent than with input controls, taken into account in the individual decision-making by fishers. As shown with the British Columbia halibut fishery and the Nova Scotia mobile gear fishery in chapters 3 and 5, individual harvesting rights create an important payoff in terms of raising the prices and revenues obtained by fishers. In part, this arises from allowing fishers to harvest over a longer period of time and from mitigation of the race to fish that allows fishers to harvest in a way that maximizes the returns per unit of fish harvested. Individual harvesting rights also allow fishers to fish in the most cost effective way without distorting optimal input combinations. By contrast to input controls and the effort creep it induces, in these incentive-based approaches to fisheries management technological change is desirable because it increases the returns from fishing, without jeopardizing sustainability.

Incentive-based approaches are not a panacea to all fishery problems and they need to be carefully designed to ensure habitat is protected as well as non-targeted species. They also need to account for overages and underages, and be designed to avoid or exacerbate problems such as highgrading. This requires adaptive management to address problems as they arise and adequate monitoring and enforcement, but this has been accomplished in a number of fisheries with individual harvesting rights, and even with multispecies fisheries (Grafton et al. 2004b). Indeed, the conclusion of the authors of a recent review of the individual quota system that has been in place in most of New Zealand's fisheries since the 1980s, is that they are an effective instrument for efficient fisheries management (Newell et al. 2005).

6.8 Future of Fisheries

Abundant evidence exists that many of the world's fisheries are managed in a way that jeopardizes sustainability and fails to deliver on their economic potential. The latest catch data from capture fisheries indicate that world catches have peaked and are now falling (Hilborn et al. 2003), and that many wild fish stocks are in decline (Pauly et al. 2002; 2003). Other approaches are clearly required as 'business as usual' is not working.

The ecosystem approach to fisheries gives a greater weight to integrated management (Garcia et al. 2003), emphasizes the importance of marine protected areas (Grafton and Kompas 2005), and is widely viewed as the alternative to past and failed management practices (Pikitch et al. 2004). The goals of the ecosystem approach are admirable – ensuring healthy marine ecosystems for the future – but unless there is explicit recognition of the incentives faced by fishers the future of fishing will not improve (Grafton et al. 2006; Hilborn et al. 2005). In other words, the key to improving outcomes in fisheries is to develop institutions that will provide the world's greatest marine predator – fishers – with the incentives to undertake sustainable practices, discourage overharvesting and prevent habitat damage.

Economics plays an important (but not exclusive) role in understanding the incentives needed to induce better behavior by fishers and to generate better management decision-making. The historical focus on fish, rather than fishers, has been observed before (Larkin 1978), but still the role of fishers has yet to be fully incorporated into management practice in many of the world's fisheries. A step in the right direction, and to 'turn the tide' (Grafton et al. 1996) in current fisheries management, is to place a premium on economic efficiency and productivity, and to make the economic viability of fisheries an important goal. As surprising as it may appear to some readers, focusing on fisher incentives and the economics of fishing is consistent with, and indeed promotes, long-term sustainability of stocks. In fact, the two countries that have gone the furthest to adopting an incentive-based approach in their fisheries – Iceland and New Zealand – are arguably the best managed.

A brighter future for fisheries involves many factors, disciplines and approaches. Our book offers just one, but often neglected, perspective – economics for fisheries management. We are convinced that the tools presented in this book, and the data and information that underlies them, provide a framework to help turnaround and improve fisheries performance to the benefits of harvesters, consumers and the environment.

References

Danielsson, A., 'Efficiency of Catch and Effort Quotas in the Presence of Risk', *Journal of Environmental Economics and Management*, 43 (2005): 20-33.

Diamond, J., *Collapse: How Societies Choose to Fail or Survive* (Camberwell, Victoria: Penguin Books, 2005).

Food and Agriculture Organization of the United Nations, *Report and Documentation of the International Workshop on the Implementation of International Fisheries Instruments and Factors of Unsustainability and Overexploitation in Fisheries* (Mauritius, 3-7 February 2003). (FAO Fisheries Report No. 700, Rome, 2004).

Garcia, S.M., Zerbi, A., Aliaume, C., Do Chi, T., and Lasserre, G., The Ecosystem approach to Fisheries. Issues, Terminology, Principles, Institutional Foundations, Implementation and Outlook (FAO Fisheries Technical Paper No. 443, Rome, 2003).

Grafton, R.Q., 'Social Capital and Fisheries Governance', *Ocean and Coastal Management* (in press).

Grafton, R.Q., Adamowicz, W., Dupont, D., Nelson, H., Hill, R.J., and Renzetti, S., *The Economics of the Environment and Natural Resources* (Malden, MA: Basil Blackwell, 2004a).

Grafton, R.Q, Arnason, R., Bjørndal T., Campbell, D., Campbell, H., Clark, C.W., Connor, R., Dupont, D.P., Hannesson, R., Hilborn, R., Kirkley, J.E., Kompas, T. Lane, D.E., Munro, G.R., Pascoe, S., Squires, D., Steinshamn, S.I., Turris, B., and Weninger, Q., 'Incentive-based Approaches to Sustainable Fisheries', *Canadian Journal of Fisheries and Aquatic Sciences*, 63 (2006): 699-710.

Grafton, R.Q. and Kompas, T., 'Uncertainty and the Active Adaptive Management of Marine Reserves', *Marine Policy* (2005): 471-9.

Grafton, R.Q., Kompas, T., and Lindenmayer, D., 'Marine Reserves with Ecological

Uncertainty', *Bulletin of Mathematical Biology*, 67 (2005b): 957-71.

Grafton, R.Q., Nelson, H.W., and Turris, B., How to Resolve the Class II Common Property Problem? The Case of British Columbia's Multi-species Groundfish Trawl Fishery (Paper presented at the Conference on Fisheries Economics and Management in Honour of Professor Gordon R. Munro, Vancouver, Canada August 5 and 6 2004), (2004b).
Available at http://www.econ.ubc.ca/munro/472grnet.pdf.

Grafton, R.Q., Sandal, L., and Steinshamn S.I., 'How to Improve the Management of Renewable resources: The Case of Canada's Northern Cod Fishery', *American Journal of Agricultural Economics*, 82 (2000b): 570-80.

Grafton, R.Q. and Silva-Echenique, J., 'How to Manage Nature? Strategies, Predator-Prey Models and Chaos,' Marine Resource Economics, 12 (1997): 127-43.

Grafton, R.Q., Squires, D., and Fox, K.J., 'Private Property and Economic Efficiency: A Study of a Common-pool Resource', *Journal of Law and Economics*, 43 (2000a): 679-713.

Grafton, R.Q., Squires, D., and Kirkley, J.E., 'Private Property Rights and Crises in World Fisheries: Turning the Tide?', *Contemporary Economic Policy*, 24 (1996): 90-9.

Hannesson, R., The Privatization of the Oceans (Cambridge: MIT Press, 2004).

Hilborn, R., Branch, T.A., Ernst, B., Magnusson, A., Minte-Vera, C.V., Scheuerell, M.D., and Valero, J.L., 'State of the World's Fisheries', *Annual Review of Environmental Resources*, 28 (2003): 359-99.

Hilborn, R., Orensanz, J.M., and Parma, A.M., 'Institutions, Incentives and the Future of Fisheries', *Philosophical Transactions of Royal Society*, 360 (2005): 47-57.

Kompas, T. and Che, T.N., 'Efficiency gains and cost reductions from individual transferable quotas: a stochastic cost frontier for the Australian south east fishery', *Journal of Productivity Analysis*, 23 (2005) 285-307.

Kompas, T., Che, T.N., and Grafton, R.Q., 'Technical Efficiency Effects of Input Controls: Evidence from Australia's Banana Prawn Industry', *Applied Economics*, 36 (2004): 1631-41.

Larkin, P.A., 'Fisheries Management — An Essay for Ecologists', *Annual Review of Ecology and Systematics*, 9 (1978): 57-73.

Ludwig, D., Hilborn, R., and Walters, C., 'Uncertainty, Resource Exploitation, and Conservation: Lessons from History', *Science*, 260 (1993): 7, 36.

Marine Conservation Biology Institute. Statement of US Scientists on Improving the Use of Science in Fisheries Management. Accessed on 5 October 2005 and available at http://mcbi.org/MFCN_statement/statement.htm.

McLoughlin, R. and Findlay, V., Implementation of Effective Fisheries Management (Presented at Outlook 05, Canberra, Australia, March 2, 2005). Available at http://www.abare.gov.au/outlook/program/day2.html.

Newell, R.G., Sanchirico, J.N. and Kerr, S. 'Fishing Quota markets', *Journal of Environmental Economics and Management*, 49 (2005): 437-62.

Organisation for Economic Co-operation and Development (OECD). Towards

Sustainable Fisheries: Economic aspects of the Management of Living Marine Resources (Paris, OECD 1997).

Pauly, D., Christensen, V., Guénette, S., Pitcher, T.J., Sumaila, U.R., Walters, C.J., Watson, R., and Zeller, D., 'Towards Sustainability in World Fisheries', *Nature*, 418 (2002): 689-95.

Pauly, D., Adler, J., Bennett, E., Christensen, V., Tyedmers, P., and Watson, R., 'The Future for Fisheries', *Science*, 302 (2003): 1359-61.

Pikitch, E.K., Santora, C., Babcock, E.A., Bakun, A., Bonfil, R., Conover, D.O., Dayton, P., Doukakis, P., Fluharty, D., Herman, B., Houde, E.D., Link, J., Livingston, P.A., Mantel, M., McAllister, M.K., Pope, J., and Sainsbury, K.J., 'Ecosystem-Based Fishery Management', *Science*, 305 (2004): 346-7.

Rose, R. and Kompas, T., *Management Options for the Australian Northern Prawn Fishery: An Economic Assessment* (ABARE Report to the Fisheries Resources Research Fund, Canberra, August, 2004).

Roughgarden, J. and Smith, F., 'Why fisheries collapse and what to do about it?' *Proceedings of the National Academy of Sciences*, 93 (1996): 5078-5083.

Sethi, G., Costello, C., Fisher, A., Hanemann, M. and Karp, L., 'Fishery management under multiple uncertainty', *Journal of Environmental Economics and Management*, 50 (2005): 300-18.

Townsend, R.E., 'Entry Restrictions in the Fishery: A Survey of the Evidence', *Land Economics*, 66 (1990): 359-78.

Walters, C.J. and Hilborn, R., 'Adaptive Control of Fishing Systems', *Journal of the Fisheries Research Board of Canada*, 33 (1976): 145-59.

Walters, C.J. and Hilborn, R., 'Ecological Optimization and Adaptive Management', *Annual Review of Ecology and Systematics*, 9 (1978): 157-88.

Glossary

Accuracy: The degree to which the value of an estimator from a sample differs from its true or population value.

Adaptive management: An approach that employs monitoring, evaluation and learning within the management process to adjust and adapt goals, strategies and tactics to new information.

Allocative efficiency: A firm or vessel is allocatively efficient when it combines inputs or factors of production in proportions that minimize the cost of producing a given level of output. In a fishery, for example, if the per unit cost of crew rises, a vessel that is allocatively efficient will substitute (to the extent possible) toward capital and technology that is labor-saving.

Artisanal fisheries: A fishery or fishing method that is relatively small scale, using traditional fishing methods usually undertaken close to shore or inshore.

Average cost: The total costs of production divided by the amount of output produced. Average costs can either rise or fall with output, or in standard cases, initially fall as output rises, reaching a minimum, then rise as output increases.

Average product: The total quantity of output divided by the total quantity of a given input. The average product of crew on a vessel is thus the ratio of harvest to the number of crew on board, measured either as the number of persons or hours spent fishing.

Base period: A period in an index number calculation to which all other periods are compared.

Biased estimate: An estimate whose expected value of the sample mean is different than the mean of the population from which the sample is drawn.

Biological growth function: A mathematical description of how a population or species changes through time. The growth of a single species of fish, for example, may depend on the existing stock of fish, its natural or intrinsic growth rate and a host of other environmental factors.

Bioeconomic model: A combination of a biological stock assessment, a stock-recruitment relationship and an economic profit equation, usually discounted over time, that determines the catch or effort level that maximizes resource rents in a fishery. Bioeconomic models range from relatively simple, deterministic, single species models to more complicated settings that allow for uncertainty and multiple species and fishing methods.

Bionomic equilibrium: A sustainable equilibrium usually associated with an open access resource, in which economic profits are zero, or where total revenues just equal the total costs of harvest.

Biomass: The total weight of all individuals in a given population.

Bycatch: Fish or other species that are caught incidentally when vessels target other species.

Capacity output: The maximum output level that can be produced given fixed inputs, the existing technology, the availability of variable inputs and the prevailing technology.

Capacity utilization: The ratio of current or observed output relative to a potential or capacity output.

Capital: Economic assets, such as durable goods, machines, fishing boats, and other plant and equipment that is used in combination with other inputs to produce goods and services.

Capital services: The flow of services or benefits provided by a capital stock (such as a vessel) over a given period of time.

Capital stuffing: In a fishery, a process where individual firms invest heavily in boat capital, gear and other equipment in order to gain a competitive advantage in the harvest of fish. The process normally results in overcapacity in the fishery, or a state where fishing capital is far larger than that needed to land a given amount of fish, usually characterized by the phrase 'too many boats chasing too few fish'.

Capital utilization: The ratio of the current or observed capital stock (such as vessel size) to the 'ideal' level of capital. Overcapitalization arises whenever capital utilization exceeds one, such that too much capital is being used relative to the ideal level. Overcapitalization is common in 'race to fish' fisheries and open access resources.

Catch per unit of effort (CPUE): A measure given by the ratio of total harvest (in weight or occasionally numbers of fish) to total fishing effort, usually measured by total days fished or total trawling hours. Catch per unit of effort (or CPUE) is often used as an imperfect measure of stock abundance. CPUE, for example, may rise not because the total stock of fish has increased, but simply because boats have entered a spatial zone or part of the fishery where the concentration of fish is larger.

Catchability coefficient: A shift parameter that accounts for the largely unexplained harvest of fish, or catches that are attributable to things other than direct fishing effort and the stock of fish. A catchability coefficient is often used in harvest equations (as functions of effort and stock) to account for changes in the technology of catching fish.

Central limit theorem: A theorem that provides assurance that with a sufficiently large enough number of observations the mean of a sample will be approximately normally distributed, and will also equal the population mean, while its variance will equal the population variance divided by the number of observations in the sample.

Census data: Data typically drawn or obtained by central governments or local government agencies from the general population on a regular basis. Variables typically surveyed are household size, income, and the level of education.

Chaos: A dynamic system characterized by extreme sensitivity to initial conditions and that generates what appears to be random behaviour, but in fact the behaviour is deterministic or determined entirely by the parameters of the system and its governing equations.

Coefficient of Variation (cv): the standard deviation of a variable divided by its mean, which gives a measure of its variability.

Cohort: A subset of a population designated by age or some other common characteristic.

Command and control regulation: Regulation that tries to modify human behavior and actions via fiat and rules.

Commercial fishery: A fishery that is characterized by large scale fishing activity, using modern fishing methods, where fishing is done largely for profit.

Common-pool resource (CPR): A resource, such as a fishery, where use is rivalrous (one person's use harms other users) and the ability to exclude users is difficult.

Common property resource: A resource over which a community or group of individuals have access to and, to some extent, are able to exclude persons from outside of the group from using.

Congestion externality: A negative externality that arises when there are too many users who impede each other and raise harvesting costs when they simultaneously try to harvest a resource.

Cross-sectional data: A data set sampled from a population at a given point in time, usually annual or seasonal.

Data envelopment analysis: A mathematical programming method used to obtain a measure of efficiency and capacity output, also referred to as DEA. DEA models are commonly used to measure capacity in a fishery, but a principal drawback is that they do not normally account for random effects.

Demersal species: Fish commonly found at or near the bottom of the sea, such as orange roughy.

Depreciation: The process by which machines and other capital goods either wear out or lose their value over time. It can be measured either as a physical or a financial loss.

Deterministic model: A model that assumes no uncertainty, as if all variables and relationships were not subject to random fluctuation.

Deterministic production frontier: A relationship between inputs and an output that represents the maximum possible amount of output that can be produced from any quantity or combination of inputs, assuming no uncertainty or randomness in production.

Diminishing returns: A process where the application of a given input (holding all other inputs constant) results in an increase in output, but at a decreasing rate. This is equivalent to stating that the marginal product of an input is positive, but falling as the amount of the input increases.

Discarding: The dumping of unwanted fish or other species, generally of low or no market value, while at sea.

Discount rate: The rate at which future income or expenditures is discounted to the present. It recognizes that a given amount of money received a year from now is worth less in terms of purchasing power today because it could be invested to yield a return in a year's hence.

Disequilibrium: A biological or mechanical process that is not in equilibrium or non-stationary, but which may be in the process of converging to an equilibrium.

Econometrics: The application of statistical and mathematical methods to the study of economic data, designed to give empirical content and verification to existing economic theories.

Economic efficiency: A state where all available inputs are used in the correct amounts and proportions to produce the maximum amount of output at the lowest cost of production. It is calculated as the product of allocative efficiency and technical efficiency.

Economic profit: The difference between total cost and total revenue, where total cost includes the opportunity cost or the value of any single input's next best alternative use. The opportunity cost of a skipper who owns a boat but does not receive a salary would thus be included in total costs, measured by the amount the skipper would receive as a hired-skipper on another boat. By contrast, accounting profit is simply the difference between total revenues and explicit cash outlays.

Economics: The science of choice, or a discipline that studies the allocation of scarce resources, with limited and alternatives uses, for the production and distribution of goods and services.

Economies of scale: A term used to describe the fall in the long run average cost of production as a result of an increase in the amount produced.

Effective fishing effort: The adjusted value of nominal fishing effort to account for technological change or the actual change in fishing power over time.

Equilibrium: A characteristic of a model that designates a resting point, or a point where all stocks or state variables are not changing over time.

Excess capacity: The difference between the output level a firm or industry could produce if production was efficient and inputs were fully utilized, and the output level actually produced for a given set of input and output prices and a given technology. In fisheries, excess capacity is often considered as undesirable because it implies too much idle capital, or wasted resources, given existing catch levels.

Expected value: A measure of the central tendency (and often mean value) of a random variable. An expected value can be unrestricted, or conditional on a set of information that may alter the value of its central tendency.

Externality: An incidental cost or benefit imposed on others from a given action.

Factors of production: The inputs used in a production process, often aggregated into the broad categories of land, labor, capital and natural resources.

Fishing effort: An aggregate measure of the amount of inputs applied to a given fishing activity, usually measured in terms of days fished, gear units or trawling hours. Standardized fishing effort is a measure of fishing effort that attempts to adjust for differences in vessel and gear characteristics.

Fixed cost: A cost of production that does not vary with the level of output such as the cost of a fishing license or the purchase price of a statutory fishing right.

GDP (Gross Domestic Product): A measure of the total value of goods and services produced by an economy in a given time interval, usually one year.

Growth rate: A measure of how a variable changes through time, usually expressed in percentage terms.

Harvesting capacity: The maximum output level (or harvest) that can be produced from a given set of inputs, or the minimum level of inputs used to produce a given harvest or output level

Heteroscedasticity: A statistical estimate where the error term does not have a constant variance. The presence of heteroscedasticity does not imply that estimated coefficients are biased, but it does affect the reliability of confidence intervals and hypotheses tests.

Highgrading: The process by which less valued fish is dumped at sea in order to maximize the value of the catch onboard.

Individual Transferable Quota (ITQ): A right to a given harvest of fish, often as a share of total allowable catch, which can be transferred (leased or sold) to other individuals or fishers.

Initial conditions: In a dynamic model, the starting value or the value of a key variable (such as the stock of fish) at time zero.

Input controls: A fisheries management strategy that attempts to control total effort by controlling the amount of inputs used in a given fishery. Typical examples of input controls include restrictions on gear length, the size of the vessel, and times and areas at which vessels are allowed to fish. Fishers often evade input controls by substituting toward unregulated or unrestricted inputs.

Input orientated measure of technical efficiency: The minimum amount of inputs required to produce a given level of output.

Input substitution: A term that describes the process by which a firm substitutes one input for another. With input controls, it is also typical for fishing vessels to substitute from regulated inputs (such as vessel size) toward unregulated inputs (such as electronic gear).

Inputs: Scarce resources, such as labor, natural resources and capital, that are used as factors of production to produce an output. In a fishery, the principal inputs or factors of production that are used to obtain a harvest of fish include crew, the vessel, gear, bait, ice, the skipper, fuel and the stock of fish.

Inverse demand function: The relationship between the quantity demanded of a particular good or service and the price that consumers are willing and able to pay for that amount.

Isocost: A line that represents the same cost of production, but that is produced with different input combinations.

Isoquant: A curve or line that represents the same level of output, but that is produced with different input combinations.

Joint production: A production process that simultaneously produces more than one output. A fishing activity, for example, may result in that catch of two or more species at once, or the catch of a target species and bycatch.

Landings data: Data on catch and vessel characteristics obtained at port, or when the harvest is delivered to port.

Logistic growth: A form of density dependent growth, or growth that depends on the stock of a resource such that, for a broad range of parameter values, an increase in stock results first in an increase in the growth rate – up to some maximum value – – followed by a decrease in growth with further increases in the resource stock, until maximum carrying capacity is reached. A common representation of logistic growth is the 'inverse U-shaped' relationship between the stock of fish and net additions (or recruitment) to the fishery.

Market-clearing price: The price in a market at which the quantity demanded of a good or service exactly equals the quantity supplied. When the quantity demanded exceeds the quantity supplied, or when there is an excess demand for the good or service, market price tends to rise through time. When quantity supplied exceeds quantity demanded the market price tends to fall.

Marginal cost: The measure of the change in the total cost of production given a marginal change in the amount of output produced.

Marginal product: The extra output or product that is obtained from a marginal change in a given input. For example, the marginal product of an additional unit of fishing effort is the additional harvest that is obtained from this increased effort.

Marginal rate of substitution: The amount of one good a person is willing to trade off for another a good.

Marginal rate of technical substitution: Also called the technical rate of substitution and is the slope of an isoquant. It represents the trade off from using marginally less of one input, and more of another input, while leaving output unchanged. Normally, we expect the marginal rate of technical substitution to diminish as we move down to the right along an isoquant.

Market-based instruments: Various approaches to resource management that rely on tradeable property rights, where the value of such rights is determined in a market context. Trades in individual transferable quotas is a good example of a market-based instrument in a fishery. In this context, trades will normally occur from high to low marginal cost producers, since low marginal cost producers can pay more for each unit of quota, thus increasing efficiency in harvest.

Market failure: Situation where the market alone, in the absence of any intervention, fails to deliver the fullest benefit to society. In fisheries market failure frequently occurs because of the presence of negative technological externalities in harvesting.

Maximum carrying capacity: The maximum number of individuals, or the biomass of a species that can be supported by the defined natural environment.

Maximum economic yield: A sustainable catch or effort level that creates the largest difference between total revenue and the total costs of fishing for the fishery as a whole. A point of maximum economic yield is often associated with a point of maximum profits in the fishery, or a catch or effort level that maximizes resource rents.

Maximum sustainable yield: A point where catch or the harvest of fish is just equal to the growth of the stock of fish (or net additions to the fish stock), which is both sustainable and at a harvest level that generates the largest possible catch. Maximum sustainable yield is generally associated with surplus-production models that show that the sustainable catch at first rises with the stock of fish, reaches a maximum, and then falls with increases in stock of fish to the point of maximum carrying capacity.

Mean: Also called the arithmetic mean, and is the sum of the values of all observations divided by the number of observations.

Median: The middle value when all observations in a sample are arranged in ascending order.

Mortality rate: The death rate of a population defined over a given period of time. In a fishery, a natural mortality rate is the rate at which fish die as a result of natural causes.

Multicollinearity: In statistical or econometric estimates, a situation where explanatory variables are highly correlated or interconnected. In such cases, it becomes difficult to disentangle the separate effects of each of the explanatory variables on the variable to be explained.

Multiple regression: A statistical or econometric approach to estimation of models where a single variable to be explained (usually termed a dependent variable) is a function of two or more explanatory (or independent) variables and an error term.

Multistage sampling: Sampling where the population is first divided into groups that are then subdivided, followed by a random sampling of the subdivisions, from which a final random sample of observations is made from the picked subdivisions.

Natural capital: A term for the value of all the environmental resources available to the economy.

Negative externality: More precisely called a negative technological externality. It represents a situation when an individual imposes costs or harm on others incidentally due to an action or activity, but does not take into account these costs in his or her decision making. For example, if one fisher catches a fish today then there is less harvest available for others to catch tomorrow, thus affecting the costs of fishing for everyone.

Nominal fishing effort: A simple measure of the total inputs applied to a fishery, usually measured as days fished or trawling hours. Nominal fishing effort is typically used in measures of catch per unit of effort.

Normal distribution: A distribution of events, observations or random variables that is bell-shaped around a given mean, with a known variance. An important property of a normal distribution is that any linear function of normally distributed variables is also normally distributed.

Observer data: Data obtained by direct observation, usually for scientific purposes.

Omitted-variable bias: A bias that alters coefficient values and other statistical measures as a result of omitting an otherwise key explanatory variable from the estimation.

Open access resource: A resource for which no effective property rights have been established, and where it is difficult if not impossible to exclude individuals from access to the resource.

Opportunity cost: An implicit cost defined as the cost or price of the next best alternative or action. For example, if the next best employment that could be obtained by an owner-operator of a fishing vessel is $30,000/year then this amount equals the opportunity cost of the skipper's labor.

Optimal control theory: A mathematical approach to dynamic problems that seeks to maximize an objective function, such as discounted profits, subject to a set of constraints and initial conditions.

Ordinary Least Squares (OLS): A statistical method for estimating the parameters (or coefficients) of a multiple linear regression model that minimizes the sum of squares of the vertical distances between actual observations and a 'best fit' line drawn through the observations.

Output controls: Controls on the harvest of a natural resource, often in a common property setting, designed to protect the resource stock, ensure sustainability and enhance profitability.

Output orientated measure of technical efficiency: A measure of actual output relative to maximum possible output given a set of inputs.

Overcapitalization: Term used for a situation when capital utilization exceeds one such that too much capital is being used relative to the ideal level.

Panel data set: Data which is obtained over time from given units of observation.

Parameter: A variable that is held constant for the moment, but that can be varied in a model context to determine the effect of its change on all other variables.

Pelagic species: Fish commonly found at or near the surface, such as tuna, pilchard and herring.

Pooled cross-sectional data: Data that combines cross-sectional observations on a sample of a population over time.

Precision: A measure of the variability of an estimator. Precision can usually be increased (variability reduced) the larger is the sample size.

Probability: The likelihood of an event occurring, typically defined as a number between zero and one. If an event occurs with certainty the probability of its occurrence is one. If the event never arises, the probability of its occurrence is zero.

Probability distribution: A mathematical formula that describes the probability of obtaining different values of a random variable.

Production frontier: A production function that shows maximum output for any given inputs, typically used in reference to either a deterministic (without randomness) or stochastic (with random effects) production frontier.

Production function: A mathematical expression that maps or transforms inputs via a production process into a single output. Constant returns to scale production functions imply that a doubling of *all* inputs would exactly double output. A decreasing (increasing) returns to scale production function implies that a doubling of all inputs results in less (more) than a doubling of output.

Productivity: Typically defined as the ratio of an output (or a collection of outputs) to an input (or a collection of inputs) in a production process.

Productivity index: A measure of productivity usually indexed to a base period or a particular observation or numéraire.

Profit: The difference between total revenues and costs.

Profit decomposition: A decomposition of changes in relative profit performance into differences in output prices and input prices, adjusted, in the case of a fishery, for their importance in the catch (outputs), fishing effort (variable inputs) and fixed inputs, such as vessel size.

Proxy variable: An observed variable that is related to an unobserved explanatory variable. For example, years of fishing experience may be used as a proxy for skipper skill.

Public bad: A good that generates disutility or harm to users and is both non-exclusive (it is not possible to prevent others from being harmed by the good) and non-rival (the use of the good by any one user does not diminish the harm incurred by others).

Public good: A good that generates benefits to users and is both non-exclusive (it is not possible to prevent others from using it) and non-rival (the use by any other user does not diminish any benefits from using the good by others). An example of a public good is a lighthouse.

Purse-seine nets: A net design that encircles schools of fish (typically near the surface) and is then closed from below trapping the fish.

Quota: A regulated amount of catch. A total quota in a fishery refers to the total allowable catch allowed in a given season while an individual quota refers to the maximum permissible harvest by an individual or vessel.

Race to fish: A term used in fisheries where boat owners compete (often by purchasing larger boats and engines) for catch. Such behavior is typical in input controlled fisheries.

Random sampling: A sampling technique where each observation is drawn at random from a population. Each draw gives an observation that is independent of previous draws such that no one observation is more likely to be selected than another.

Random variable: A variable that can take on a certain value with a defined probability. A discrete random variable is a variable that assumes only a particular finite or countable set of values (such as the number of boats in a fishery). A continuous random variable can assume any value within a certain range.

Resilience: The ability of a stock or resource to return to a former state following a shock. Usually it is measured by the speed or time it takes to return to its former state.

Response accuracy: Error that arises in a survey where the respondent inadvertently gives an incorrect or inaccurate answer.

Response bias: Error that arises in a survey where the respondent deliberately gives an incorrect or inaccurate answer, possibly to influence decisions that might be based on the survey results.

Resource rent: Loosely defined as the net return after all the economic costs (including opportunity costs) of production have been deducted from total revenues.

Returns to scale: The proportional change in output from a given proportional change in all inputs. Constant returns to scale implies that a doubling of *all* inputs would exactly double output. Decreasing (increasing) returns to scale implies that a doubling of all inputs results in less (more) than a doubling of output.

Sample bias: Bias or differences that arise between the mean of a sample and the population from which it is drawn because the sample is not representative of the population.

Scale efficiency: Production at a level of output, and with an amount of plant or capital equipment, that would maximize profits if the firm were economically efficient.

Social discount rate: The discount rate at which society chooses to discount future costs and benefits. This social discount rate would normally be lower than a private or individual discount rate.

Standard deviation: A measure of the spread of the distribution of a random variable. The standard deviation of a random variable is the square root of its variance, which summarizes the expected difference from the mean or central tendency of a random variable.

Statistical inference: The procedures by which observed data are used to draw conclusions about the population from which the data came, or about the unknown process that generated the data.

Steady state equilibrium: A resting point where all variables are growing at the same rate. If the growth rate of all variables is zero, the equilibrium point is stationary.

Stochastic process: A sequence designating the value of a random variable or a group of random variables through time. A serially independent stochastic process is one where the realization or exact value of a random variable at a point in time is independent of the value of the same random variable in previous periods. Serially correlated stochastic processes imply that the value of a random variable at a point in time in part depends on its past values.

Stochastic production frontier: The maximum amount of output produced for any given set of inputs, allowing for random variation.

Stochastic production frontier analysis: A combination of a stochastic production frontier and an inefficiency model, forming an assessment of the variables that determine output and the firm-specific variables that result in differences in efficiency.

Stock assessment: A study of the population of a species in a given fishery to determine the likely stock-recruitment relationship and to estimate its biomass.

Stock-recruitment relationship: A construction that tries to model and measure the relationship between the stock of fish (or some comparable biomass measure) and new recruits to the fishery.

Stratified sampling: Sampling where a population is divided into groups, called strata, and random samples are then chosen from the strata.

Surplus-production model: Frequently referred to as the Schaefer model in fisheries and is a mapping between the stock of fish and the net additions (or recruitment) to the fishery. Basic surplus-production models often assume no age distribution of fish and no natural mortality, thus establishing a hypothetical relationship showing the growth in the stock of fish as a function of the stock of fish. This relationship is usually presented as an 'inverse U-shaped' relationship, where the left-hand intercept designates a virgin fishery and the right-hand intercept establishes maximum carrying capacity. Between these two extremes lies maximum sustainable yield.

Sustainable yields: A harvest or catch that theoretically equals the net additions to the stock of fish, thus leaving the stock of fish unchanged through time. For a standard surplus-production model, for example, there are an infinite number of sustainable yields corresponding to each value of stock.

Systematic sampling. Sampling which involves listing the population and then picking every tenth or every hundredth, etc. observation.

Technical efficiency: A measure of how close a firm is to producing the maximum amount of output from given inputs. Unlike allocative efficiency, technical efficiency is not concerned with the proportions in which inputs are used, but at a given proportion, measures the extent to which a given amount of inputs results in maximum output.

Time-series data: Data where each variable has a time index. Average harvest in a fishery for every year for a fifty-year period is an example of time series data.

Total Allowable Catch (TAC): A fisheries management strategy that determines the maximum amount of fish (by individual species) that can be harvested in a fishery over a given period of time, usually defined as the fishing season.

Total revenue: The total receipts from fishing, or the price of fish times the quantity of fish landed.

Tragedy of the commons: A term associated with the biological and economic overexploitation of a common-pool resource because of open access.

Transitional path: The path taken to a steady state equilibrium. To move from one steady state sustainable yield to another requires a transitional period where harvest may be much lower (or higher depending on the direction of adjustment) than in the final resting point or steady state.

Truncated normal distribution: A normal distribution that takes on only zero or positive values, frequently used as the distribution for the technical inefficiency term in a stochastic production frontier analysis.

Type I error: Probability of rejecting a null hypothesis that is true.

Variable costs: Costs that vary with output. In fisheries, fuel and bait expenditures are variable costs.

Variance: A measure of the expected difference from the mean or central tendency of a random variable, or a measure of the spread of the distribution of a random variable.

Virgin biomass: The hypothetical value of biomass before any fishing has taken place.

Index